Hacking Math Class with Python

Exploring Math Through Computer Programming

By Peter Farrell

ISBN-10: 1508656940

ISBN-13: 978-1508656944

Email: peter@farrellpolymath.com

Website: www.farrellpolymath.com

Contents

1. Introduction to Programming

Tools

In the popular video game Minecraft, you're "spawned" into a random world with no instructions. You need to make yourself a shelter and some tools. First you gather wood blocks to make a crafting table and wood tools. With the wood tools you can gather stone to craft harder stone tools. With stone tools you can mine iron ore to craft even harder iron tools.

| Wooden Block and pickaxe | Stone Block and Pickaxe | Iron Ore and Pickaxe |

It goes on like this: you use the tools you have to make bigger, more effective tools. Once you've made stone tools, there's no reason to go back to wooden tools!

Math is like that, too. You start out learning the tools of arithmetic and build on those to create algebraic tools and so on. For centuries this "tool-making" was simply symbolic or psychological, but with the availability of free computer applications, we can craft tools that will help us explore higher and higher math topics.

I'm not talking about using calculators or computers to avoid doing math. I'm talking about writing programs to avoid excessive repetition of tasks. These can be simple two-line functions to return the average of two numbers, or much more complicated programs to draw graphs or model 3D situations.

Python

The folks who developed the Python programming language have given us a bunch of pre-loaded tools we can use to create our personal tools:

- Loops
- Conditionals
- Variables
- Functions
- Lists
- Classes

The difference between this and Minecraft is that there are plenty of real-world applications of learning to program, even if you start by making turtles walk around.

Another difference is that there are programmers and developers out there right now creating even newer tools you can use if you find them, download them and read their documentation. Once you master the Python tools listed above you can simply "import" other packages and use all the tools they've made for you!

Installing Python:

If you already have Python on your computer, skip this section. The **Raspberry Pi** already has Python pre-loaded on it, including IDLE, the basic text editor we'll be using to write programs.

Python works on any operating system, so you can download it for free at python.org. I'm going to use Python 3. Usually you download an installer file which you double-click to run and just follow the directions to install it. Pretty easy!

Running Python:

Double-click the desktop icon for Python or find it in the dropdown menus under Programming:

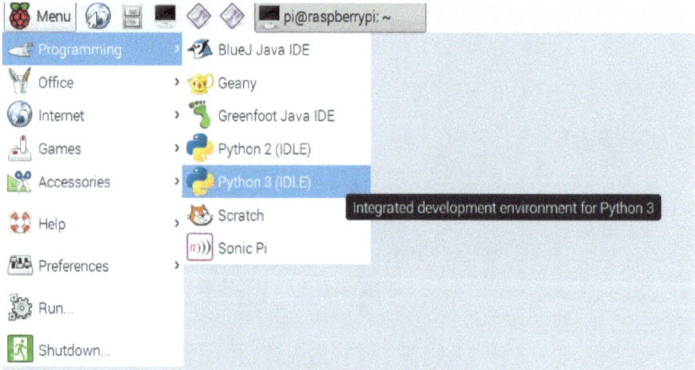

You get an interactive "shell."

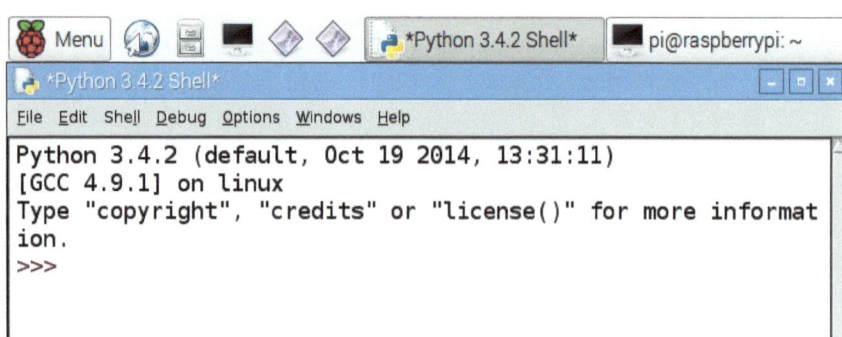

Notice the little carrots (>>>). They're called the "prompt." You can use the shell as a calculator by typing in an expression and pressing ENTER:

```
>>> 12 + 45
57
>>> 97 - 53
44
```

Multiplication is an asterisk.

```
>>> 65 * 33
2145
```

Two asterisks mean an exponent:

```
>>> 5**3
125
```

Forward slash means division:

```
>>> 100/12
8.333333333333334
```

The first thing most people learn is how to **print** some text: in Python 3 it couldn't be easier. Type:

```
>>> print("hello world!")
```

and press Enter. It prints whatever you put in the parentheses! Congratulations. You just ran your first line of code!

The `Turtle` **Module**

In the 1960s MIT was interested in creating a programming language that would be easy to use, even for children. This culminated in the Logo programming language, as discussed in Seymour Papert's book *Mindstorms*. There are many applications (like NetLogo) that you program in Logo. Scratch and Snap are drag-and-drop versions of Logo programming. The creators of Python thought the Logo turtles were so important they included a turtle module in Python.

A **module** in Python is a file with code in it. The "turtle module" is a file called "**turtle.py**" (all Python files need the .py extension), and it's somewhere in the Python folders when you download the language. turtle.py contains all the functions you can use to make your turtles move around and make cool graphics!

Using the Turtle Module

Importing the turtle module is easy. If you type

```
import turtle
```

at the beginning of a program, you can call the functions in the turtle module by typing "turtle" before every command, like "`turtle.fd(10)`". You can shorten this by importing the module this way:

```
import turtle as t
```

Then you just have to type t before every command, like "`t.fd(10).`"

I prefer even less typing. **My recommendation** is to import the turtle module this way:

```
from turtle import *
```

and then you don't have to type anything else before your turtle commands.

The great thing about Python is that it gives you pop-up messages to show you've typed the command in the right way. For example, in the Python shell, when you type this code, you see the message asking you for the distance you want the turtle to move forward.

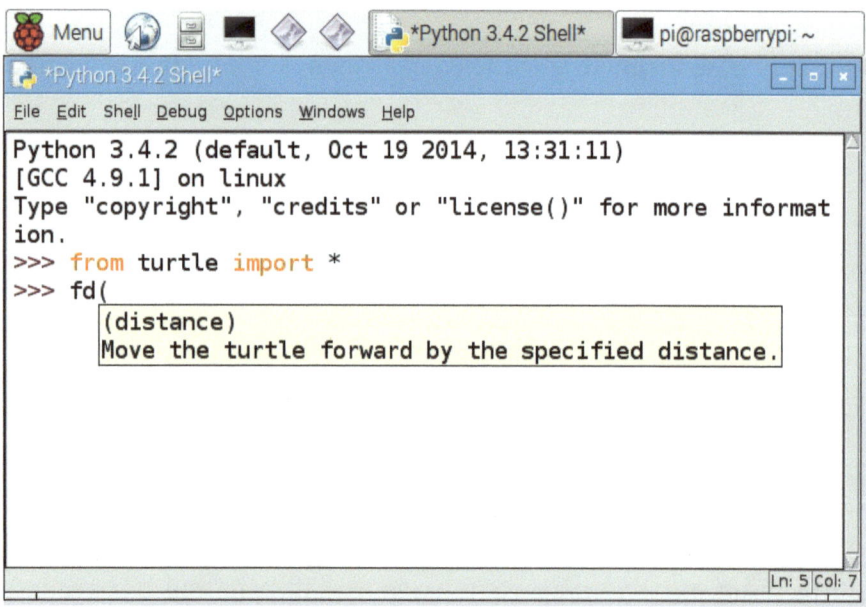

Type in a value and press ENTER.
```
>>> fd(100)
```

The Turtle Graphics window should pop up with your turtle and the path it leaves from walking forward 100 steps.

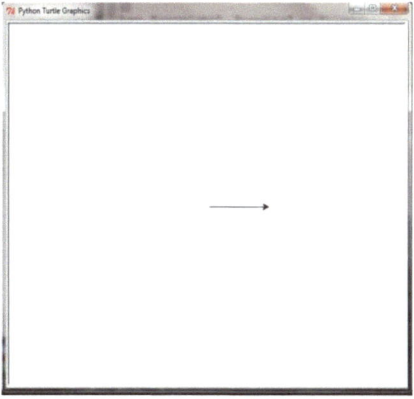

If you want to clear your turtle graphics, simply type

```
>>> clear()
```

into the shell and the turtle path will be erased (but the turtle will still be standing there).

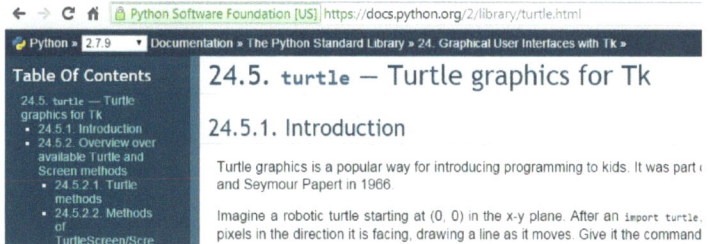

All the commands you'll need can be found in the Python.org docs, under the turtle module. Just Google "Python turtle" and it should be the first one that pops up.

Making Turtles Move

If we're going to do any serious programming we have to leave the interactive shell and create permanent Python files we can save.

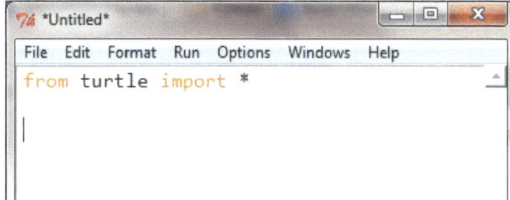

Open a **new window** by clicking "File → New Window" or by pressing the CTRL key and N at the same time. <u>Notice there's no prompt or carrots.</u> This is called a **module window** and we can save modules and run them again later or share them with other people.

The first time you run a file (using the "Run" dropdown at the top of IDLE and selecting "Run Module" or by pressing F5), you'll be prompted to save the file. Name it something like "turtle1.py" because there is already a "turtle.py." After that, you simply have to press F5 to run a program. And you can store all the programs we write for our turtles in "turtle1.py."

How to Make a Square

Here's the code to make your turtle draw a square:

```python
from turtle import *

fd(100)
rt(90)
fd(100)
rt(90)
fd(100)
rt(90)
fd(100)
```

Press F5 or click "Run" and you'll see a square in the turtle window:

Using Loops

You could do the same thing with less typing by using a **loop**, a very important tool that will show up in nearly every computer program. Every language has its own way of making the computer repeat a block of code a certain number of times, and in Python it's

```python
for i in range(4):
```

to make it repeat 4 times. i is a variable you can use in the code (and we will use it soon) but for right now all you need to know is that's how to create a repeat loop.

To repeat a block of code 5 times, the code is

```python
for i in range(5):
```

To repeat a block of code 10 times, it's

```python
for i in range(10):
```

and so on.

When you type a colon and press ENTER, Python automatically indents the next line (or lines) for you. The indented block will be what's repeated four times. The indenting is really important! If your code isn't indented correctly, it won't work properly. So here's our loop:

```python
for i in range(4):
    fd(100)
    rt(90)
```

The outcome (when we run it) is a square.

Congratulations! Your first Python program! Play around with the length to make different squares, and change the angle to make other shapes.

Making a Triangle

Changing your code from drawing a square to drawing a triangle is a good exercise in geometry. Obviously instead of repeating the code 4 times you repeat it 3 times for a triangle, since it only has 3 sides. But how many degrees do you turn to make a triangle? The simple answer is 60, but that's the internal angle. If you adapt your square code like this:

```python
for i in range(3):
    fd(100)
    rt(60)
```

here's what it looks like in the Turtle Graphics Window:

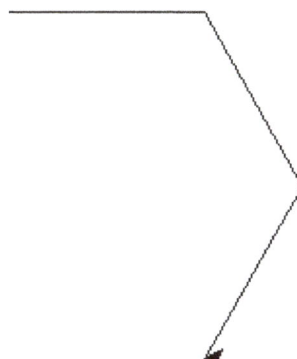

What the turtle does is turn the **external** angle, not the internal angle. So to draw a triangle, you have to make the turtle turn 120 degrees each time.

```
for i in range(3):
    fd(100)
    rt(120)
```

Defining Functions

A good way to organize your code is to name a set of commands, so you can call them easily:

```
def square():
    for i in range(4):
        fd(100)
        rt(90)
```
"def" means **define** the function

Notice that everything is indented inside the `square` function, and everything inside the loop is indented again. The good news is IDLE automatically indents for you after you type a colon.

Now we can just type
```
>>> square()
```

in the shell and the turtle will draw a square. We can even call the `square` function inside another function, like this:

```
def squareThing():
    color('green')
    for i in range(20):
        square()
        rt(10)
```

> **Functions** are a <u>very</u> important programming tool you'll use in every program, no matter what language you use.

all the "square" code will automatically be run in the new function. In this case, the turtle will make a square, then turn right 10 degrees, and repeat that 20 times. Run it and type

```
>>> squareThing()
```

into the Python shell and press Enter. The figure should look like this:

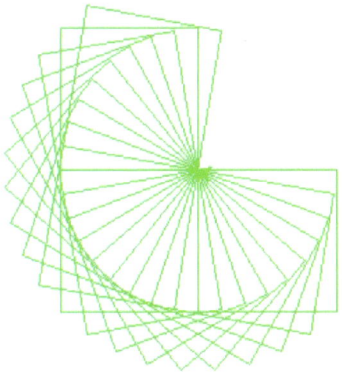

How would you make a circle?

```python
def circle():
    for i in range(360):
        fd(1)
        rt(1)
```

Using Variables

Variables are another very useful tool. You replace numbers with a letter or a word and then you can change every instance of the variable at once.

Program: square with side `length`

Change your "square" function to:

``python
def square(length):
    for i in range(4):
        fd(length)
        rt(90)
``
	Now whatever value is put in the parentheses when you type in the square function will be put wherever you see "length" in the function.

Now you can draw squares of any length you want:

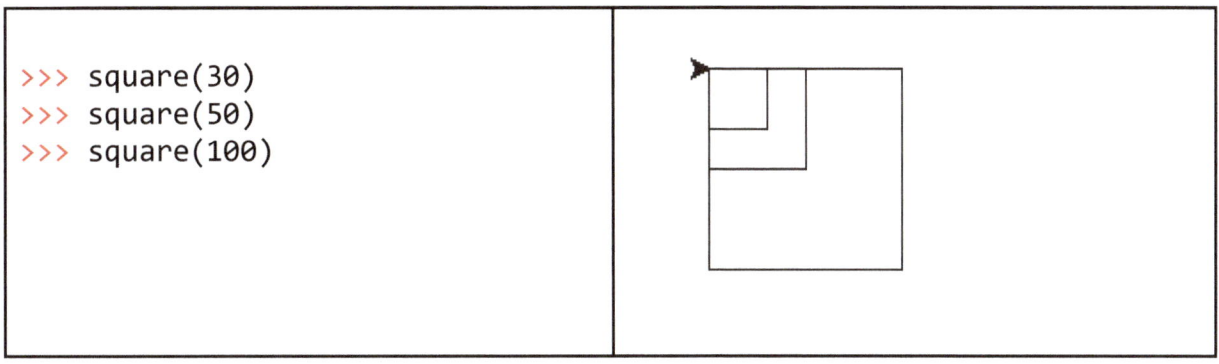

```python
>>> square(30)
>>> square(50)
>>> square(100)
```

Whatever you put in the parentheses will be saved to the "length" variable and applied wherever that variable is in the function. Unfortunately, if you don't specify a length from now on, you'll get an error message because the program is looking for a length.

```
>>> square()
Traceback (most recent call last):
  File "<pyshell#3>", line 1, in <module>
    square()
TypeError: square() takes exactly 1 argument (0 given)
```

I read the last line of the errors first. It's saying the square function expects to be told the length. In programming lingo, it "takes 1 argument." We'll write functions that take more than one argument later. There's a way to tell Python, "if a user doesn't tell you a length, just use 100," for example. Just change the parameter (what's in the parentheses) to:

```
def square(length = 100):
```

Now if you specify a length, it will draw a square with that length. If you leave the parentheses empty, then the length will default to 100.

```
>>> clear()
>>> square()
>>> square(80)
```

Program: Spiral

```
def spiral(times = 30):
    '''draws a spiral of squares.'''
    color('blue')
    length = 10
    for i in range(times):
        square(length)
        rt(5)
        length = length + 5

spiral()
```

Polygon Function

What about a polygon with more sides, like a pentagon or hexagon? We have to think like a turtle. We start at the center, facing a certain direction. We turn a certain number of times and end up at the same spot, facing our original direction. How many degrees have we turned? 360. So each turn will be 360 divided by the number of turns. The number of degrees we turn for a pentagon will be 360 / 5, which is 72. So our pentagon program will look like this:

```
for i in range(5):
    fd(100)
    rt(72)
```

To draw a regular polygon with n sides, how many degrees should the turtle turn?

$$360/n$$

Telling the turtle how to draw a polygon depends on how many sides you want. This is called the "parameter" of a function. Let's return to our pentagon program and let the user choose how many sides they want. I'm changing the name to "polygon":

```
def polygon(sides):
    for i in range(sides):
        fd(100)
        rt(360/sides)
```

And at the prompt:
```
>>> polygon(7)
```

Using subroutines

Program: Flower

How would you make a complicated design like this flower?

First you learn to make a petal by putting together two quartercircles. Do you remember how to make a circle?

```
def circle():
    for i in range(360):
        fd(1)
        rt(1)
```

How would you change that code to make a quartercircle? Think about it.

```
def quartercircle():
    for i in range(90):
        fd(1)
        rt(1)
```

Then you can make a "petal" function composed of two quartercircles. Experiment with the angle between the quartercicles before peeking at the code!

```
def petal():
    for i in range(2):
        quartercircle()
        rt(90)
```

The flowerhead is made up of twelve petals. So you would make a flowerhead function and then finally put them all together into a flower function. The great news is all you have to do is execute the flower function and all the subroutines will run automatically!

```
def flowerhead():
    for i in range(12):
        petal()
        rt(30)

def flower():
    speed(0)
    color('green') # green stem and leaf
    pensize(3)     # for thicker lines
    setpos(0,-200) # start lower on screen
```

```
setheading(90) # face straight up
fd(100)        # draw the stem
petal()        # draw a leaf on the stem
fd(150)        # finish the stem
color('pink') # pink flower petals
flowerhead()   # draw the flowerhead
```

By the way, did you notice I wrote little notes in the program? It's a good idea to write "**comments**" to explain your code, and that's why every programming language has a syntax for comments. In Python, it's the pound sign or hashtag (#). Anything on the line after the hashtag will be ignored by the computer. It's only for humans to read!

```
>>> print(3)
3
>>> #print(3)
```
(Nothing was printed)

```
'''You can also comment out more than one line by using
three quotes or three single-quotes around the whole region!'''
```

```
def square():
    '''This draws a square.'''
    for i in range(4):
        fd(100)
        rt(90)
```
Add a description of your function in triple quotes.

```
>>> square(
        ()
        This draws a square.
```

The comment shows up to explain the function when you type square(
Try it!

By now you've worked with 3 of the major tools of Programming:

Loops
Variables
Functions

In a future section we'll be doing much more with turtle graphics including graphing, fractals and differential equations!

Let's take a break from turtles for a moment and learn some more tools to help explore the world of math.

Loops and Printing

You might not always be working with turtles, but very often you'll be printing something, even to check your programs. Here's how to print.

Program: simple print

```
>>> print("Hello World!")
Hello World!
```

In Python 2, you don't have to use parentheses for printing:

```
>>> print "Hello World!"
Hello World!
```

Program: print strings using variables

<pre>def print3():
    for i in range(10):
        print(i)</pre>
	<pre>>>> print3()
	0
	1
	2
	3
	4
	5
	6
	7
	8
	9</pre>

It printed 10 numbers, but it starts at 0. If you want to print starting with 1, you need to start the range with 1, and end it with 11:

<pre>def print3():
    for i in range(1,11):
        print(i, end = '')</pre>
Notice the "end" code prevents a line break.
	<pre>>>> print3()
	12345678910</pre>

Program: print strings using loops

| ```python
def print3():
    for i in range(4):
        print(2*i + 1)
``` | ```
>>> print3()
1
3
5
7
``` |

Program: print numbers up to `maxnum`

| ```python
def print1(maxnum):
    for i in range(maxnum):
        print(i, end = '')
``` | ```
>>> print1(11)
012345678910
``` |

Remember Python (and computers in general) starts counting with 0, not 1. To print the numbers from 1 to maxnum, including maxnum, you'd have to change the range:

| ```python
def print1(maxnum):
    for i in range(1,maxnum+1):
        print(i)
``` | ```
>>> print1(10)
1
2
3
4
5
6
7
8
9
10
``` |

While loops

There's also a way to make the program keep doing something while a certain condition is true. For example, keep printing numbers while they're less than or equal to maxnum. You could use a **while loop**:

| ```python
def printUpTo(maxnum):
    num = 0
    while num <= maxnum:
        print(num,end = ' ')
        num += 1
``` | ```
>>> printUpTo(10)
0 1 2 3 4 5 6 7 8 9 10
``` |

The Tools So Far

This chapter has given you an introduction to the tools you can use to create your own tools using Python:

Loops
Variables
Functions
Turtle Graphics
Print Statements
Comments

You'll see all these tools in future chapters, so get used to using them!

Turtle Exercises:

Using the above tools, you should be able to create simple shapes, then use those shapes to make even more complicated shapes. Try these:

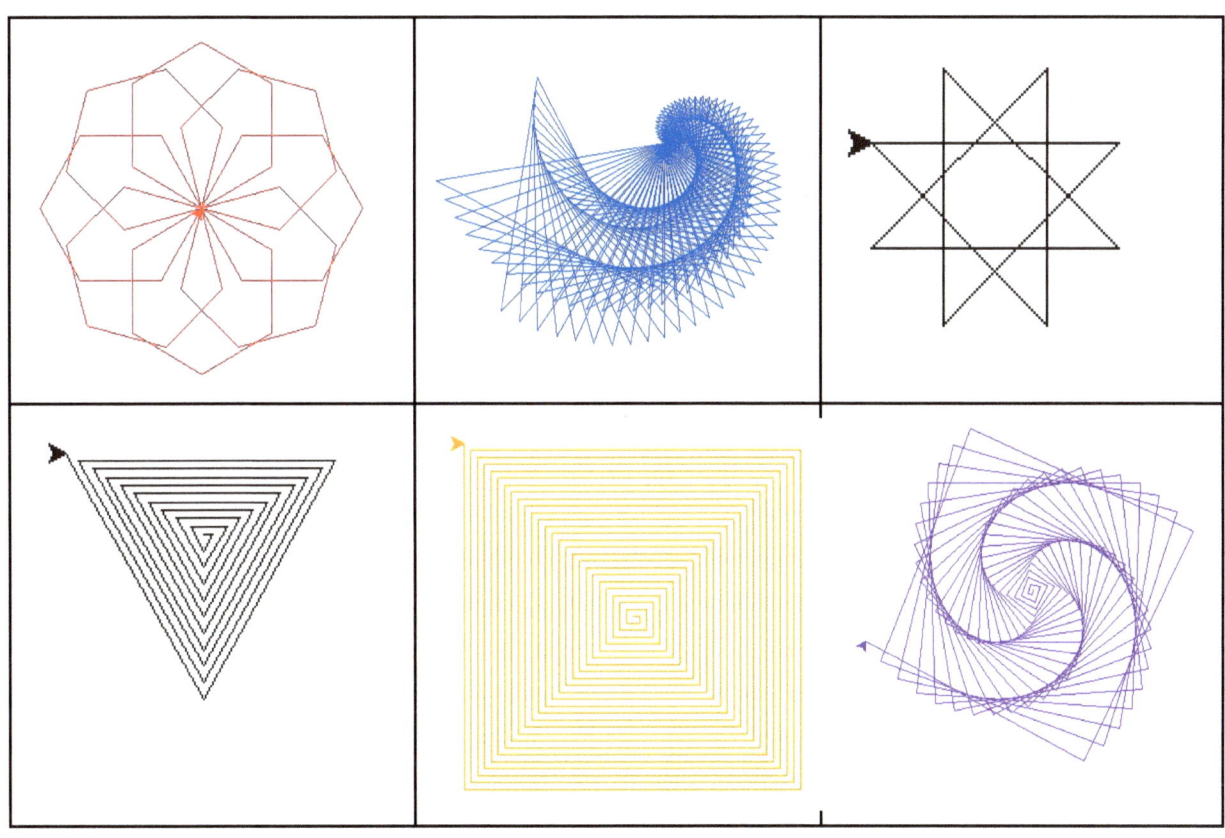

2. Arithmetic

Functions

So far we've used functions to draw things and print numbers. The real use of functions in math is to **return** values. Think input and output:

```
def function(input):
    return output
```

That's not a real function, of course. Let's create some real ones that do something to our input and return the output. Notice you don't need parentheses around what you want to return:

```
def double(x):
    return 2 * x

def oneLess(x):
    return x - 1

def specialSquare(x):
    return x**2 + 3
```

It's up to you what you name a function. You can't use spaces or most punctuation but you can use capital letters and underscores. It's a good idea to make your function names descriptive.

Let's use the functions one by one. I'll put 4 into the first function and put the output into the second function and so on.

```
>>> double(4)
8
```

We get an output of 8. Let's put that into the next function:

```
>>> oneLess(8)
7
```

The output is 7. Let's put that into the third function:
```
>>> specialSquare(7)
52
```

The final output is 52. We could do all that in one line. Here's how to automatically make the output of one function the input of the next function. The innermost parentheses are evaluated first.

```
>>> specialSquare(oneLess(double(4)))
52
```

That might seem like a lot of parentheses to keep track of right now, but I hope it gives you a taste of what you can do with Python functions!

Program: Average

Here's a useful tool to craft. We'll need it in Geometry for finding midpoints.

```def average(a,b):``` ``` return(a+b)/2```	```>>> average(10,15)``` ```12.5```

Program: Convert Fahrenheit to Celsius

Another common task is converting numbers according to a formula. Here are two functions that illustrate conversion tools we'll see later on.

```def celsToFahr(celsius):```
```   '''Converts Celsius to Fahrenheit'''```
```   print((9/5)*celsius + 32, 'F')```
 	
```def fahrToCels(fahrenheit):```
```   '''Converts Fahrenheit to Celsius'''```
```   print((5/9)*(fahrenheit - 32), 'C')```
	```>>> celsToFahr(20)```
	```68.0 F```
	```>>> fahrToCels(212)```
	```100.0 C```

Conditionals and Input

There's a number-guessing game in which you try to guess the number I'm thinking of. If you don't guess it I have to tell you "higher" or "lower." How many guesses would it take to guess correctly if my number is between 1 and 10? 1 and 100? 1 and 1000? You get the idea. Before we can find out we need to learn a few more programming tools.

Conditionals

Another common task when programming is telling the program to check whether something is true. Every programming language has a syntax for "True" and "False" (in Python they're capitalized); these statements are called "Booleans." To check whether two things are equal, you need two equal signs:

```
>>> 2 = 2
SyntaxError: can't assign to literal
```

```
>>> 2 == 2
True
```

Often you'll be checking whether a number is greater or less than another:
```
>>> 5 > 4
True
>>> 2 * 5 > 100
False
```

Conditionals check whether a statement is true, and if so, do something. It's sometimes called an "if-then" statement, but in Python, there's no "then." For example:

```
>>> if name == 'Peter':
        print('That's my name, too!')
```

The Number Game

Let's start a new module called numbergame.py and start by saying hello:

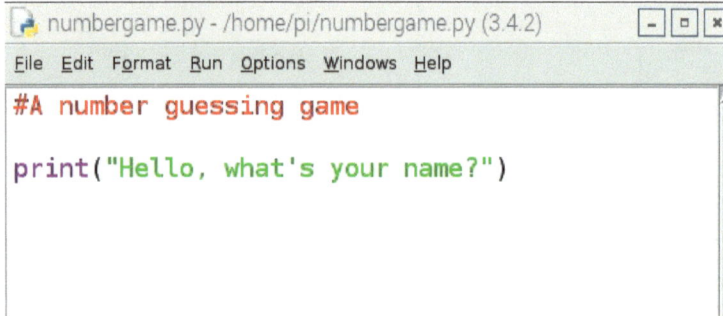

Notice that the first line of code is a **comment**. When you run this program, you'll get

```
>>>
Hello, what's your name?
>>>
```

But if you try to enter your name, you'll get an error message:

```
Hello, what's your name?
>>> Peter
Traceback (most recent call last):
  File "<pyshell#20>", line 1, in <module>
    Peter
```

24

```
NameError: name 'Peter' is not defined
>>>
```

You might see this error message in the future, if Python doesn't recognize a variable or some other code. Right now Python doesn't know what to do with what I typed. It printed the line it was supposed to, and that's all we told it to do.

User Input

In order to take in user input, we have to use Python's "input" function. In Python 2, it's "raw_input." Change the code to this:

```
name = input("What's your name? ")
```

Now whatever the user types in, it will save that string to a variable called `name`.
```
What's your name? Peter
>>> print(name)
Peter
```

Input differences in Python 2 & 3

Python 2:	Python 3:
raw_input("...")	input("...")

Add this line to your code:

```
print("Hi,", name)
```

Now, whatever the user types in, the program will greet them.
```
What's your name? Peter
Hi, Peter
```

To play our number guessing game, we could choose a number for the user to guess, but what if they want to play again and again? It'll be easier to just let the computer choose a random number for us. That will require the random module.

The random module

You can use **the random module** to generate random numbers and choose randomly from a range or a list. Here's how you use it in our number guessing game.

```
import random
```

```
number = random.randint(1,10)
```

It will save a random number between 1 and 10 to a variable called "number." Now you can use **Conditional statements** to respond differently according to the input the user enters:

```
guess = int(input("What's your first guess? "))
if guess == number:  #notice the double equals sign!
    print("That's it!")
elif guess < number: # "elif" means "otherwise"
    guess = int(input("Nope. Higher."))
elif guess > number:
    guess = int(input("Nope. Lower."))
```

You need the "int" before the "input" because **user input is always in the form of a string, not a number**. "int" changes the string to an integer.

"elif" means "otherwise, if…" and you can have numerous "elif"'s in a conditional, if there are a lot of possible conditions.

Remember: checking for equality requires **double equals signs** (==). A single equals sign is how you assign a value to a variable.

Put together some loops and you have your number guessing game:

Program: Number Guessing
```
import random

def numbergame(upperlimit):
    '''Plays a number guessing game.'''
    guesses = 1
    win = 0
    name = input("Hi! What's your name? ")
    print("Hi,",name)
    print("Let's play a game.")
    while True: #infinite loop
        print("I'm thinking of a number between 1 and ", upperlimit)
        guess = int(input("What's your first guess? "))
        number = random.randint(1,upperlimit)
        playing = True
        while playing:
            if guess == number:
```

```
                print("That's it!")
                print("You guessed it in", guesses, "guesses!")
                win = 1
                break
            elif guess < number:
                guess = int(input("Nope. Higher."))
            elif guess > number:
                guess = int(input("Nope. Lower."))
            else: #if the input isn't a number
                guess = int(input("Error."))
            guesses += 1

        replay = input("Play again? y/n")
        while replay != "y" and replay != "n":
            replay = int(input("Error."))
        if replay == "n":
            break
        elif replay =='y':
            guesses = 1     #reset the number of guesses
            win = 0         #reset the win to 0

numbergame(100)
```

Here's the output:

```
Hi! What's your name? Peter
Hi, Peter
Let's play a game.
I'm thinking of a number between 1 and 100
What's your first guess? 50
Nope. Higher.75
Nope. Higher.87
Nope. Higher.92
Nope. Lower.90
Nope. Higher.91
That's it!
You guessed it in 6 guesses!
Play again? y/n
```

Back to our original question. How many tries would it take in the above program to guess a number between 1 and 100? What if the number is between 1 and 1000? Make a conjecture and try it out!

Lists

We learned to use a variable to save a number's value, but how would you save a bunch of values? A very useful way to group items together is using **lists**. Every programming language has lists, and in Python they're easy to use. Group the items inside square brackets, separated by commas:

```
>>> a = [1,2,3]
>>> print(a)
[1, 2, 3]
>>> type(a) # checks a's type
<type 'list'>
```

You can add lists (called "concatenation") to get a list with all the items combined.

```
>>> b = ['hello','there']
>>> a + b
[1, 2, 3, 'hello', 'there']
```

You can add items and remove them using the append() and remove() functions:

```
>>> a.append(10)
>>> print(a)
[1, 2, 3, 10]
>>> a.remove(2)
>>> a
[1, 3, 10]
```

Items have a position in a list called an **index**. Here's how to call the first item in a list:

```
>>> a = [6,3,5,2]
>>> a[0]
6
```

In most programming languages, numbering starts at 0. The second item in the list will have index 1.

```
>>> a = [6,3,5,2]
>>> a[1]
3
```

List indices can be negative, too. That means "starting from the last item in the list."

```
>>> a = [6,3,5,2]
>>> a[-1]
2
>>> a[-2]
5
```

Program: Median

In math class you're often asked to find the median of a list of numbers. It's easy to put them in a list and sort them using Python's "sort()" function but a little tricky to choose which one is the median. If the length of the list is an odd number n, the median is the term with the index that's the whole number part of n/2. The Python syntax for "just the whole number part of a division" is "//".

```python
def median(listA):
    '''Returns the median of a list of numbers.'''
    #first put the list in order:
    listA.sort()
    n = len(listA) #n is the length of the list
    #If there's an even number of items in the list:
    if n % 2 == 0:
        #The median is the average of the middle two
        return average(listA[n//2-1],listA[n//2])
    else:
        #The median is the middle one
        return listA[n//2]
```

Program: Factoring a number

A common task in arithmetic is factoring a number. This program will come in handy when we do algebra.

```python
def factors(number):
    '''Returns a list of factors of a number.'''
    factor_list = [1] #1 is definitely a factor
    for i in range(2, number+1): #try every number up to 'number'
        if number % i == 0: #if number mod i is zero
            factor_list.append(i) #add it to the list
    return factor_list
```

The Modulo Operator

Haven't come across "mod" before? The "Modulo Operator" sounds scarier than it really is. In Python (and some other languages, too) the percent sign (%) is used to divide two numbers and only return the **remainder**. We're using it exclusively to check if the remainder is zero.

```
>>> 10%5
0
>>> 12%5
2
>>> 20%7
6
```

The remainder when you divide 10 by 5 is zero. When you divide 12 by 5 the remainder is 2, and when you divide 20 by 7 the remainder is 6. Let's check to see if our `factors` function works:

```
>>> factors(24)
[1, 2, 3, 4, 6, 8, 12, 24]
```

Yep, those are all the factors of 24.

Generating the Fibonacci Sequence

Here's a good illustration of the use of negative list indices. The first two Fibonacci numbers are 1 and 1. To get the next Fibonacci number, you add the previous two together.

```
def fibo(n):
    '''Returns n Fibonacci numbers.'''
    fibos = [1,1] #The first two Fibonaccis
    for i in range(n-2):
        fibos.append(fibos[-1] + fibos[-2])
    return fibos
```

Now to get the next Fibonacci number the program just adds the **last two in the list**. To get 10 Fibonacci numbers, enter

```
>>> fibo(10)
[1, 1, 2, 3, 5, 8, 13, 21, 34, 55]
```

Lists are necessary if you want to make a random choice from a number of items. First import the random module:

```
>>> import random
>>> a = [6,3,5,2]
```

```
>>> random.choice(a)
2
>>> random.choice(a)
3
>>> random.choice(a)
5
```

We'll make extensive use of lists in all the math topics that follow. We'll get a lot of practice declaring lists, adding to them, using their index numbers and iterating over their elements. They're a very useful tool!

The tools so far

In this chapter we've learned about some new Python tools:	We've made a few useful tools of our own:

Conditionals
 Lists
 Sort
 Modulo (%)
 Random Module
	`average`
	 `celsToFahr`
	 `fahrToCels`
	 `median`
	 `factors`
	 `fibo`

We'll use these in future chapters to make even better tools. The good news is *we'll never have to code them again*! But if you open a new file and try to use your factors function, for example, Python will give you an error message which ends like this:

```
NameError: name 'factors' is not defined
```

So let's save all the arithmetic functions we created in a file called **arithmetic.py**. If we need any of those functions, we can import the arithmetic module the same way we've imported the turtle module and random module. You're a Python developer already! In the next chapter we can start a new file called **algebra.py** for example, and as long as we put it in the same directory/folder as arithmetic.py we can use the `factors` function by typing this:

```
from arithmetic import factors
```

```
a = factors(99)
print(a)
```

Run it and the output will be
```
[1, 3, 9, 11, 33, 99]
```

Do you see how major this is? It means you don't mind putting in a lot of effort to craft a tool because you'll always have it in your tool chest. In the next chapter we'll create tools for drawing a grid and graphing functions, not to mention solving equations.

Arithmetic Exercises (solutions on page 140):

Use the functions you created in this chapter to answer these questions:

1. What is the average of 225 and 723?

2. What is the average of 1,412 and 36,877?

3. Convert 212 degrees Fahrenheit to Celsius.

4. My friend was telling me about the weather in Brazil, saying, "It's wonderful! It's 25 degrees!" I realized he meant Celsius. What's that in Fahrenheit?

5. Find the median of these numbers: 18,12,11.5,14,9,21,8,15,3,25,10,18,6

6. Find the median of these numbers: 76,59,64,23,11,98,56,77,91,89,48,101,55,37

7. What is the 20th Fibonacci number?

3. Algebra

Solving Equations ax + b = cx + d

Algebra is like one big number guessing game.

"I'm thinking of a number. If you multiply it by 2 and add 5, you get 21."	$2x + 5 = 21$

For a long time in Algebra class, there are hundreds of equations to solve that can be put in the form $ax + b = cx + d$. Sometimes there's no x term on one side, meaning the coefficient is zero.

$$3x - 5 = 22$$
$$4x - 12 = 2x - 9$$
$$\frac{1}{2}x + \frac{2}{3} = \frac{1}{5}x + \frac{7}{8}$$

Using a little Algebra, you can solve the general form of the equation and that will help you solve all equations of that form.

$$ax + b = cx + d$$
$$ax - cx = d - b$$
$$x(a - c) = d - b$$
$$x = \frac{d - b}{a - c}$$

Now we can write a function to take the coefficients of the equation you want to solve and return that value of x:

```python
def equation(a,b,c,d):
    '''Returns the solution of an equation
    of the form ax + b = cx + d'''
    return (d - b)/(a - c)
```

See how I translated the algebra solution to the output of the function? To solve our first equation $3x - 5 = 22$, enter the coefficients, with c being 0:

```python
>>> equation(3,-5,0,22)
9.0
```

That's true: 3(9) - 5 = 22. How about the second equation? (4x - 12 = 2x - 9)

```
>>> equation(4,-12,2,-9)
1.5
```

Yes, 4(1.5) - 12 = -6 and 2(1.5) - 9 = -6.

To solve the equation

$$2x - 8 = 4x + 5$$

Enter this:
```
>>> equation(2,-8,4,5)
-6.5
```

Using the Pythagorean Theorem

Here's a very useful program to solve for the third side of a right triangle using the Pythagorean Theorem:

```
from math import sqrt

def pythag(leg1,leg2):
    '''solves for hypotenuse'''
    return sqrt(leg1**2 + leg2**2)

def pythag2(leg1,hyp):
    '''solves for other leg
    of right triangle'''
    return sqrt(hyp**2 - leg1**2)
```

```
>>> pythag(5,12)
13.0
>>> pythag2(24,25)
7.0
```

Solving Higher-Degree Equations

Algebra is where Python starts to really get useful. At some point in Algebra, the book starts using "f(x) =" instead of "y =". Mathematical functions are like machines where you put a number in (the input), the machine does something to it, then returns a number (the output).

We've already seen how easy this is in Python. We've made functions that take parameters like "length" and "number" like:

```
def square(length):
    for i in range(4):
        fd(length)
        rt(90)
```

and

```
def equation(a,b,c,d):
    return (d - b)/(a - c)
```

We already know how to take in input and return output. So when an Algebra book gives us a problem like this one:

Let $f(x) = x^2 + 5x + 6$. Find $f(2)$ and $f(10)$.

We can define the function f(x) like any Python function, input numbers and get the output instantaneously:

```
def f(x):
    return x**2 + 5*x + 6

>>> f(2)
20
>>> f(10)
156
```

Program: Factoring Polynomials

Some algebra textbooks make you factor a lot of polynomials. Here's how to make a computer do it by brute force.

You know the expression $ax^2 + bx + c$ (if it's factorable) must split up into the form
$$(dx + e)(fx + g)$$

d and **f** are factors of **a** and **e** and **g** are factors of **c**. All you have to do is go through all the factors of **a** and all the factors of **c**, and see if d * g + e * f equals b. An annoying job for a human but the computer doesn't mind all that repetition. We can't just import our factors function for this one, because we need to include the negative factors. Here's the updated function:

```
def factors(n):
    '''returns a list of the positive and
    negative factors of a number'''
    factorList = []
    n = abs(n) #absolute value of the number
    for i in range(1,n+1): #all the numbers from 1 to n
        if n % i == 0: #if n is divisible by i
            factorList.append(i) #i is a factor
```

```
        factorList.append(-i) # so is -i
    return factorList
```

Now factoring the polynomial is just a matter of plugging in the factors of a and c and seeing which combination adds up to b.

```
def factorPoly(a,b,c):
    '''factors a polynomial in the form
    ax**2 + b*x + c'''
    afactors = factors(a) #Get the factors of a
    cfactors = factors(c)   # and c
    for afactor in afactors: #try all the factors of a
        for cfactor in cfactors: #and c
            #see which combination adds up to b
            if afactor * c/cfactor + a/afactor * cfactor == b:
                print('(',afactor,"x +",cfactor,")(",
                    int(a/afactor),"x +",int(c/cfactor),")")
                return
```

Here's how to factor the polynomial $6x^2 - x - 15$:
```
>>> factorPoly(6,-1,-15)
( 2 x + 3 )( 3 x + -5 )
```

Program: Quadratic Formula

But factoring polynomials is of limited value because not all quadratics are factorable. There's a better way to solve equations involving an x^2 term: the quadratic formula. It'll give you solutions that are whole numbers or decimals, real numbers or imaginary!

```
def quad(a,b,c):
    '''Returns the solutions of an equation
    of the form a*x**2 + b*x + c = 0'''
    x1 = (-b + sqrt(b**2 - 4*a*c))/(2*a)
    x2 = (-b - sqrt(b**2 - 4*a*c))/(2*a)
    return x1,x2
```

If you need to solve the quadratic equation $x^2 + 3x - 10 = 0$, you would enter 1 for a, 3 for b and -10 for c this way:

```
>>> quad(1,3,-10)
```

```
(2.0, -5.0)
```

That means the values for x that make the equation true are 2 and -5. To check them, you plug in those values for x: $2^2 + 3(2) - 10 = 0$ and $(-5)^2 + 3(-5) - 10 = 0$

With a little more coding the quadratic formula can give you solutions whether they're whole numbers, fractions, decimals, irrationals, even imaginary numbers. It all depends on whether the "discriminant," $b^2 - 4ac$, is positive or negative.

```python
from math import sqrt

def quad(a,b,c):
    '''returns solutions of equations of the
    form a*x**2 + b*x + c = 0'''
    discriminant = b**2 - 4*a*c
    if discriminant >= 0: #real solutions
        x1 = (-b + sqrt(discriminant))/(2*a) #first solution
        x2 = (-b - sqrt(discriminant))/(2*a) #second solution
        return x1, x2
    else: #imaginary solutions
        real_part = -b/(2*a)
        imaginary_part = sqrt(-discriminant)/(2*a)
        print(real_part,"+",str(imaginary_part)+"i")
        print(real_part,"-",str(imaginary_part)+"i")
```

Now it will solve the quadratic even if there are only imaginary solutions.

```
>>> quad(1,3,10)
-1.5 + 2.7838821814150108i
-1.5 - 2.7838821814150108i
```

Higher order solutions (for equations with x^3 or x^4 and so on) require other methods.

Brute Force
You could just plug in every number you can think of. Actually, this is pretty easy for a computer. To solve the equation $6x^3 + 31x^2 + 3x - 10 = 0$, for example, you could just have the computer plug in every number between -100 and 100 for x and if it equals zero, print it out:

```python
def plug():
    for x in range(-100,100):
        if 6*x**3 + 31*x**2 + 3*x - 10 == 0:
            print("One solution is", x)
```

```
    print("Done plugging.")

>>> plug()
One solution is -5
Done plugging.
```

So the other solutions to that equation are not integers. You could change the program to plug in decimals, too.

```
def plug():
    x = -100
    while x <= 100:
        if 6*x**3 + 31*x**2 + 3*x - 10 == 0:
            print("One solution is", x)
        x += 0.5 #Will go up in half-unit steps
    print("Done plugging.")

>>> plug()
One solution is -5.0
One solution is 0.5
Done plugging.
```

But is there a better way? There is a name for a method of solving equations by showing every possible input and output of a function within a certain range and it's called **graphing**.

Major Math Tool: Create Your Own Grapher

What better tool is there for exploring functions than a graph? There are many graphing calculators and programs out there (some are free!) but many math teachers don't allow students to use graphers. What if the student <u>made</u> the grapher tool? We can use the `turtle` module to create graphs of lines and curves. First you have to define how to draw a grid.

```
from turtle import *
def setup():
    speed(0)          #Sets turtle speed to fastest
    setworldcoordinates(-5,-5,5,5) #lower left and upper right corners
    setpos(0,0)       #sets turtle's position to (0,0)
    clear()           #Clears its trails
    setheading(0)     #Faces the right of the screen
    pd()              #Puts its pen down to draw
    color("black")    #Sets its color to black
```

```
for i in range(4):    #"Do this four times"
    setpos(0,0)       #Centers itself again
    fd(5)             #Go forward 5 steps
    rt(90)            #turn right 90 degrees
pu()                  #Sets its pen in up position
```

If you type "setup()" in the shell you'll see this:

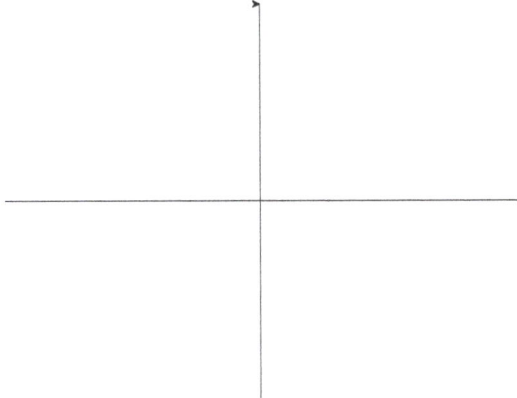

Now you have to define the function for the turtle to draw.

```
def f(x):              #This defines the function f(x)
    return 2*x + 3     #The line y = 2x + 3

def graph(f):       #The function f(x) has to be defined
    speed(0)        #Sets speed to the fastest
    color("black")  #Sets its color
    setpos(-6,f(-6))     #Sets its position to the left edge
    pd()                 #pen down
    setheading(0)        #Face right
    x = xcor()   #set a variable, x, to the x-coordinate of the turtle
    while xcor() <= 6:   #Do this until the x-coordinate is more than 6
        x += 0.01        #make x go up a tiny bit
        goto(x,f(x))     #Go to the next point on the graph
```

You've defined the functions, but you haven't told the program to run those functions yet. You can type commands in the shell or add this to the end of the program:

```
#Execute these functions automatically on "run"
setup()
graph(f)
```

Your output should be a line:

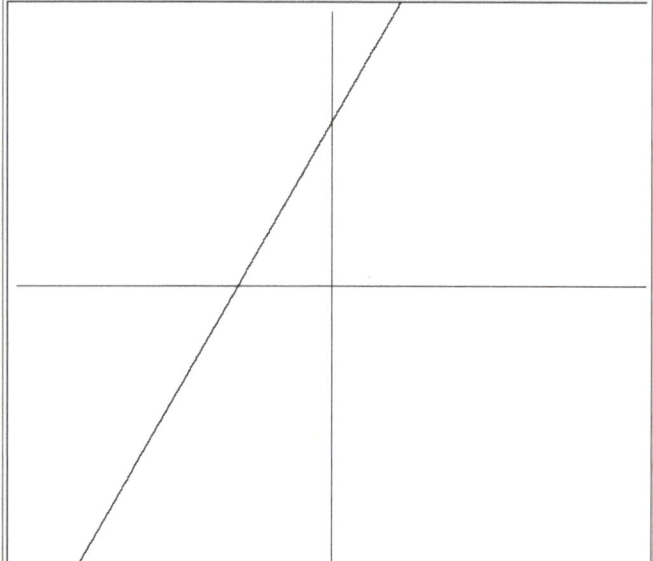

The great thing about having this tool is it will come in handy for all your math subjects from Algebra to Differential Equations!

If you add ticks to the axes, then you can graph a function and maybe the solution will be easy to see. Here's the setup program for a grapher with ticks:

```python
#Grapher with tick marks at whole numbers
from turtle import *

def mark(): #to make a tick mark on an axis
    rt(90)   #turn right 90 degrees
    fd(0.1) #go forward a little bit
    bk(0.2) #go back twice as far. The tick is drawn.
    fd(0.1) #go forward to the axis
    lt(90)   #turn left to continue drawing the axis

def setup():
    speed(0) #Sets turtle speed to fastest
    setworldcoordinates(-6,-5,6,5) #lower left and upper right corners
    setpos(0,0)      #sets turtle's position to (0,0)
    clear()          #Clears its trails
    setheading(0)  . #Faces the right of the screen
    pd()             #Puts its pen down to draw
    color("black")  #Sets its color to black
    for i in range(4):  #"Do this four times"
        setpos(0,0)     #Centers itself again
```

```
    for i in range(6): #Make 6 tick marks
        fd(1)
        mark()
    rt(90)
pu()                #Sets its pen in up position
```

Run this and you'll get a pretty graph.

We'll use our grapher to solve our equation $6x^3 + 31x^2 + 3x - 10 = 0$. First, graph the function $f(x) = 6x^3 + 31x^2 + 3x - 10$.

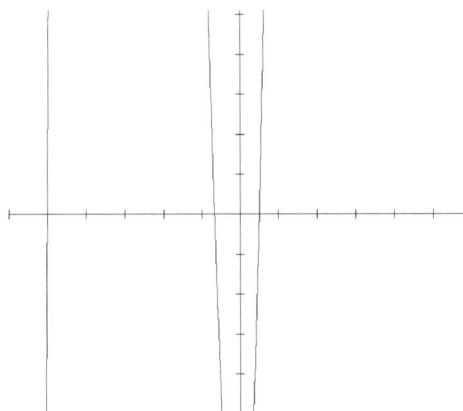

Change the "def f(x)" line of code:

```
def f(x):#This defines the function f(x)
    return 6*x**3 + 31*x**2 + 3*x - 10
```

and run it. The graph is on the left. We knew about the solutions at x = -5 and at x = 0.5 and now we can tell there's a solution between -1 and 0.

Rational Roots

We can narrow down all the possible rational roots (whole numbers or fractions) by dividing all the factors of the constant term (in this case, -10) by all the factors of the coefficient of the highest degree of x (in this case, 6). This is the list, positives and negatives:

$\pm 1, \pm 2, \pm 5, \pm 10, \pm \frac{1}{2}, \pm \frac{1}{3}, \pm \frac{1}{6}, \pm \frac{2}{3}, \pm 5/3, \pm \frac{5}{6}, \pm 10/3$

It used to be a whole lot of work to plug in a couple of dozen numbers, but that was before computers. Now we can just write a program to generate all those roots and plug them in for us. First we need the **factors** function (which we saved in our *arithmetic.py* file), which returns a list of all the factors of a number:

```
from arithmetic import factors
```

We can use this function to generate all the possible rational roots of our equation. Here's the function:

```
def rationalRoots(a,b):
    '''Returns all the possible rational roots of
```

```
a polynomial with first coefficient a and
constant b'''
roots = [] #create a list to store the roots
numerators = factors(b) # generate a list of factors of b
denominators = factors(a) # generate a list of factors of a
for numerator in numerators: #loop through the numerator list
    for denominator in denominators: #and the denominator list
        roots.append(numerator / denominator)
        roots.append(-numerator / denominator)
return roots
```

Now if I put in 6 for a and 10 for b:

```
>>> rationalRoots(6,10)
[1.0, -1.0, 0.5, -0.5, 0.3333333333333333, -0.3333333333333333,
0.16666666666666666, -0.16666666666666666, 2.0, -2.0, 1.0, -1.0,
0.6666666666666666, -0.6666666666666666, 0.3333333333333333, -
0.3333333333333333, 5.0, -5.0, 2.5, -2.5, 1.6666666666666667, -
1.6666666666666667, 0.8333333333333334, -0.8333333333333334, 10.0, -
10.0, 5.0, -5.0, 3.3333333333333335, -3.3333333333333335,
1.6666666666666667, -1.6666666666666667]
```

We already have a plug function, so I can just change it a little to plug in all those values.

```
def plug2(factorlist):
    for x in factorlist:
        if 6*x**3 + 31*x**2 + 3*x - 10 == 0:
            print("One solution is", x)
    print("Done plugging.")
```

Execute the plug function on the factor list:

```
>>> plug2(rationalRoots(6,10))
One solution is 0.5
One solution is -0.666666666666
One solution is -5.0
One solution is -5.0
Done plugging.
```

The root x = -5 came up twice (-5/1 and -10/2). **Now we know our three solutions: x = -5, x = -⅔ and x = ½.**

Now we can use the tools we just created to easily generate all the rational solutions to **any** polynomial equation. For example, if we wanted to solve the equation

$$48x^5 - 44x^4 - 884x^3 + 321x^2 + 3143x + 980 = 0.$$

What a monstrosity! But the worst part is typing it all out:

```
def h(x): #define the function so we can use "graph(h)"
        return 48*x**5 - 44*x**4 - 884*x**3 + 321*x**2 + 3143*x +980

def plug3():
    '''plugs every item in a list of factors into h(x)'''
    for x in rationalRoots(48,980):
        if h(x) == 0:
            print("One solution is", x)
    print("Done plugging.")
```

Now the program does all the work instantly:
```
>>> plug3()
One solution is 4.0
One solution is 2.5
One solution is -3.5
One solution is -1.75
One solution is 2.5
One solution is -3.5
One solution is -1.75
One solution is 2.5
One solution is -3.5
One solution is -1.75
Done plugging.
```

There's some repetition here; there are four rational solutions above.
x = -3.5, -1.75, 2.5 and 4.

Here's the graph:

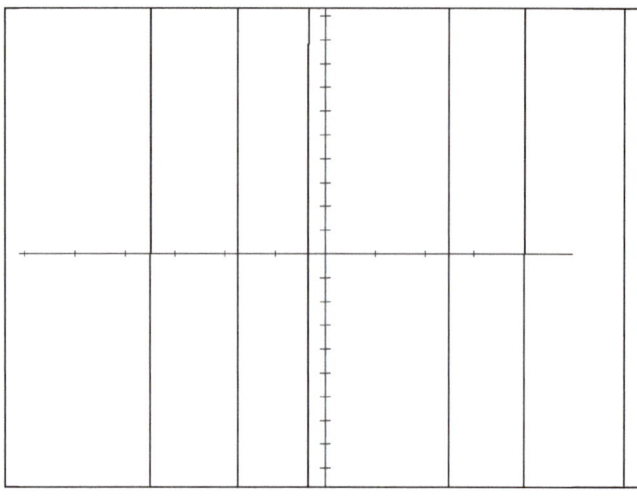

It looks like there's another root between -1 and 0. This method fails if there's a repeating decimal, like -0.33333… because if you plug that decimal into the equation you don't get zero. You get something tiny like a trillionth, but technically it's not equal to zero so our function threw it out. But there is a way to use the roots we know to get more roots.

Synthetic Division

Synthetic division is a way to do long division on polynomials. It's time consuming to do by hand, but it's easy to write a program to make the computer do it. Because we know $x = 4$ is a root of $48x^5 - 44x^4 - 884x^3 + 321x^2 + 3143x + 980$, we know we can write it as

$$(x - 4)(???) = 48x^5 - 44x^4 - 884x^3 + 321x^2 + 3143x + 980$$

"(???)" is a fourth-degree polynomial. We can divide the polynomial $48x^5 - 44x^4 - 884x^3 + 321x^2 + 3143x + 980$ by $x - 4$ to find out what it is.

```python
def synthDiv(divisor,dividend):
    '''divides a polynomial by a constant and returns a lower-degree
    polynomial. Enter divisor as a constant: (x - 3) is 3
    Enter dividend as a list of coefficients:
    x**2 - 5*x + 6 becomes [1,-5,6]'''
    quotient = [] #empty list for coefficients of quotient
    row2 = [0]    #start the second row
    for i in range(len(dividend)):
        quotient.append(dividend[i]+row2[i]) #add the ith column
        row2.append(divisor*quotient[i]) #put the new number in row 2
    print(quotient)
```

And here's how to enter it:
```
>>> synthDiv(4,[48,-44,-884,321,3143,980])
[48, 148, -292, -847, -245, 0]
```

Those are the coefficients of the new polynomial. It's
$48x^4 + 148x^3 - 292x^2 - 847x - 245$. The last number is the remainder of the division, which is 0.

So we've factored

$$48x^5 - 44x^4 - 884x^3 + 321x^2 + 3143x + 980$$

into

$$(x - 4)(48x^4 + 148x^3 - 292x^2 - 847x - 245)$$

We can keep doing this with all the other factors we know, like 2.5. Copying and pasting the factor list:

```
>>> synthDiv(2.5,[48, 148, -292, -847, -245])
[48, 268.0, 378.0, 98.0, 0.0]
```

Now we've factored our polynomial further into

$$(x - 4)(x - 2.5)(48x^3 + 268x^2 + 378x + 98)$$

Using the factor x = -3.5:

```
>>> synthDiv(-3.5,[48, 268.0, 378.0, 98.0])
[48, 100.0, 28.0, 0.0]
```

Now our factored polynomial becomes:
$$(x - 4)(x - 2.5)(x + 3.5)(48x^2 + 100x + 28)$$

We can put the coefficients in the x^2 expression into the Quadratic Formula to find our last two roots:

```
>>> quad(48,100,28)
-0.333333333333 -1.75
```

We already knew x = -1.75 is a solution, and now we know the last one is x = `-0.33333333`.

Synthetic Division can help us find all the rational roots of a polynomial, no matter how ugly it is! It can even help us find irrational roots, because we used the quadratic formula to find the last two roots.

Use this program to find all the solutions to the equation
$$8x^6 - 206x^5 + 1325x^4 + 1273x^3 - 14980x^2 - 17825x - 4875 = 0$$

Hint: four solutions are rational, and two are irrational.

Exploring Prime Numbers

Program: IsPrime()

Here's a good exploration for learning to use loops and conditionals. It tests whether a number is prime. You have to divide by 2, then 3, and so on up to which number? The number minus one? Half the number?

Remember the modulo or "mod" operator. Its symbol is the percent sign (%). The remainder when you divide 10 by 3 is:

```
>>> 10 % 3
1
```

This allows us to check whether a number is divisible by another one. Let's create a "divisible" function:

```python
def divisible(a,b):
    '''Returns True if a is divisible by b'''
    return a % b == 0
```

We can use this function inside an "isPrime" function. To test whether 61 is prime, we just divide 61 by every number less than 61:

```python
def isPrime(n):
    for i in range(2,n): #every number from 2 to n - 1
        if divisible(n,i): #if n is divisible by i
            return False    #n is not Prime (and stop)
    return True      #if it hasn't stopped, n is Prime
```

Here's how you check:
```
>>> isPrime(61)
True
```

But what if it isn't? You need to factor the number if it isn't Prime:
```python
def isPrime2(n):
    '''Returns "True" if n is Prime'''
    for i in range(2,n):
        if divisible(n,i):
            print (n,"=",i,"x",n/i)
            return
    return True
```

```
>>> isPrime2(161)
161 = 7 x 23.0
```

How many numbers do we really have to divide by? For example, we check the number 101. The number we divide by (i) starts off smaller than n/i, but somewhere it gets bigger. Can you tell where?

```
n        i        n/i
101      1        101.0
101      2        50.5
101      3        33.6666666667
101      4        25.25
101      5        20.2
101      6        16.8333333333
101      7        14.4285714286
101      8        12.625
101      9        11.2222222222
101      10       10.1
101      11       9.18181818182
101      12       8.41666666667
101      13       7.76923076923
```

It's between 10 and 11. This means there's no way there could be a factor of 101 bigger than 11 because it would have to be multiplied by a number smaller than 11. **And we've checked all those numbers already.** What's this magic point in between 10 and 11? **The square root of 101.**

Change your code to:

```
from math import sqrt
def isPrime3(n):
    m = sqrt(n)
    for i in range(2,int(m) + 1): #range has to be integers
        if divisible(n,i):
            print(n,'=',i,'x',n/i)
            return
    return True
```

Can you modify this code to print out <u>every prime number</u> up to "n"?

Here's how to print out a list of n primes. We'll use the original "isPrime" function and create a "primeList" function:

```python
def primeList(n):
    prime_list = []
    num = 2
    while len(prime_list) < n:
        if isPrime(num):
            prime_list.append(num)
        num += 1
    print(prime_list)
```

To get a list of 10 Primes:
```python
>>> primeList(10)
[2, 3, 5, 7, 11, 13, 17, 19, 23, 29]
```

Binary Numbers

How can we convert a decimal number to binary?

27	11011
decimal	binary

The place values in decimal are ones or 10^0, tens or 10^1, hundreds (10^2) and so on. First we need to find out the largest power of 2 we need to work with. In the case of 27, the we check to see the lowest exponent of 2 that's smaller than 27. 2^5 is bigger so 2^4 (16) is the highest power of 2 we're dealing with. Here's how you check for that in Python.

First loop through the powers of two until the power of two is bigger than the decimal number. Then take away 1 from the exponent:

```python
number = 27
exponent = 0
binary_number = 0
while number >= 2**exponent:
    exponent += 1
exponent -= 1
```

"exponent" is now 4, so that's 5 times we have to check powers of 2. If "number" is bigger, we take away that power of 2 and add that power of 10 so there's a 1 in that column.

48

if 27 is bigger than 2^4, take away 16 (number is now 11) and add 10^4

if 11 is bigger than 2^3, take away 8 (number is now 3) and add 10^3

and so on.

```python
for i in range(exponent + 1):
    if number - 2**exponent > -1:
        binary_number += 10**exponent
        number -= 2**exponent
    exponent -= 1
```

And when the loop is done, print out the binary number. Here's the whole code:

Program: Binary converter

```python
def binary(number):
    '''Converts decimal number to binary'''
    exponent = 0      #We're dealing with exponents of 2
    binary_number = 0               #The binary form of the number
    while number >= 2**exponent: #Finds the lowest power of 2
        exponent += 1               #the number is less than
    exponent -= 1
    for i in range(exponent + 1):
        if number - 2**exponent > -1: #If number contains power of 2
            binary_number += 10**exponent #Add that power of 10
            number -= 2**exponent   #Take away that power of 2 from
                                    #number
        exponent -= 1               #Next lower exponent
    return binary_number
```

What's the number 30 in binary?

```python
>>> binary(30)
11110
```

Yes, because $16(1) + 8(1) + 4(1) + 2(1) = 30$

The tools so far

We've made quite a few useful tools in this chapter!

```
equation
pythag
factors
factorPoly
graph
plug
rationalRoots
synthDiv
isPrime
binary
```

Save them to a file called **algebra.py** and we'll be able to easily import them for future use.

Algebra Exercises (Answers on page 140)

In problems #1 - 8, solve the equation for x.

1. $3x - 5 = 34$.

2. $7x + 25 = 2x - 20$

3. $x^2 - 13x + 40 = 0$

4. $5x^2 - 25x - 29 = 0$

5. $12x^3 + 68x^2 - 115x - 21 = 0$

6. $105x^4 + 326x^3 - 369x^2 + 34x + 24 = 0$

7. $315x^6 - 709x^5 + 1870x^4 - 1473x^3 - 5743x^2 + 3976x = 588$

8. $378x^7 - 4737x^6 + 5380x^5 + 6548x^4 - 6439x^3 - 4190x^2 + 1256x + 704 = 0$

9. What is the (base 10) number 44 in binary?

10. Is 1,000,001 (a million and 1) a prime number? If not, factor it.

4. Geometry

Expressing Points as Lists

In Geometry we often deal with points expressed as "ordered pairs" like (2, 5). We can express these pairs as lists like [2,5] or tuples, which use parentheses: (2,5). Either way, in this chapter and the chapters following, you'll see list indices to refer to the x-coordinate and y-coordinate of a point. For example, let's call (2, 5) "point A." Here's how we make point A into a list in Python:

```
>>> pointA = [2,5]
>>> type(pointA)
<type 'list'>
>>> pointA[0]
2
>>> pointA[1]
5
```

This works when you use tuples, too. Tuples are like lists, but they use parentheses:

```
>>> pointB = (2,5)
>>> type(pointB)
<type 'tuple'>
>>> pointB[0]
2
>>> pointB[1]
5
```

We really just want access to the x- and y-coordinates of our points, so it doesn't really matter which form we use. But now when you see the index notation, you'll know what's going on.

Program: Drawing a circle with a given center and radius

Let's make a turtle draw a circle for future use. Start a file called geometry.py. We saved all our code from the Algebra chapter in a file called algebra.py. Now we can import functions from it, like setup and graph.

```
from turtle import *
from algebra import setup, graph
from math import pi
```

```
def drawCircle(center,radius):
    '''draws a circle with given center and radius'''
    speed(0)
    pu()
    goto(center)      #First go to the center
    shape('circle') #set turtle to circle shape
    clone()           #Leave a point at the center
    fd(radius)        #move to the circumference
    lt(90)            #turn to move along the edge
    arc_length = 2*pi*radius/360 #length of segment on arc
    pd()              #put pen down to draw
    for i in range(360): #"do this 360 times"
        fd(arc_length)  #draw the segment
        lt(1)            #turn a little
    ht()               #hide turtle
```

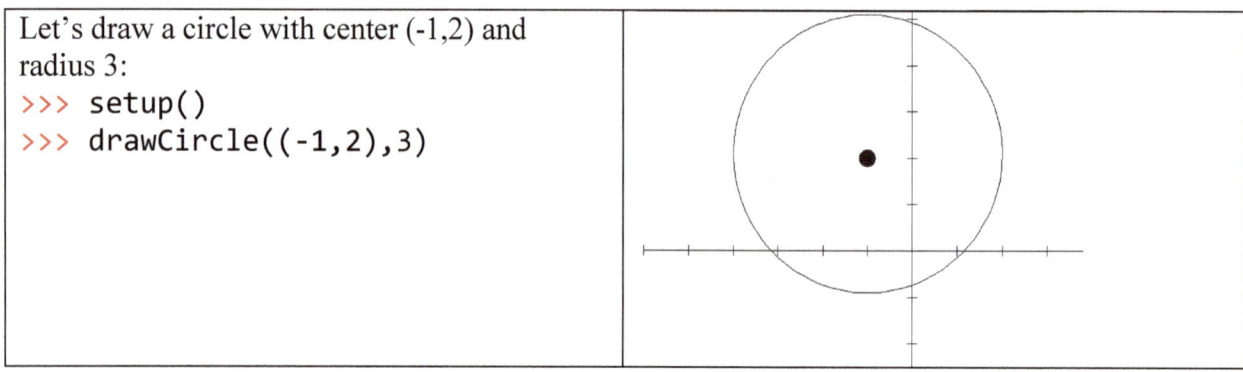

Let's draw a circle with center (-1,2) and radius 3:
```
>>> setup()
>>> drawCircle((-1,2),3)
```

Program: Finding the line between two points

To find the line between two points (a, b) and (c, d), first you find the slope by dividing the vertical change by the horizontal change:

$$m = \frac{d - b}{c - a}$$

Then you plug that slope (and either x and y pair) into the equation of a line y = mx + b and solve for b, the y-intercept: b = y – mx. Here's how to do that in Python:

```
def line2points(point1,point2):
    '''Returns the slope and y-intercept of
    the line between two points'''
    slope = (point2[1] - point1[1])/(point2[0] - point1[0])
    y_int = point1[1] - slope*point1[0]
    return slope, y_int
```

Here's how to find the equation of the line between (2,5) and (-3,-5).

```
>>> line2points((2,5),(-3,-5))
(2.0, 1.0)
```

The output is the slope and y-intercept. That means the line between (2,5) and (-3,-5) is
$y = 2x + 1$. To check it, let's graph the line and see if it passes through both points:

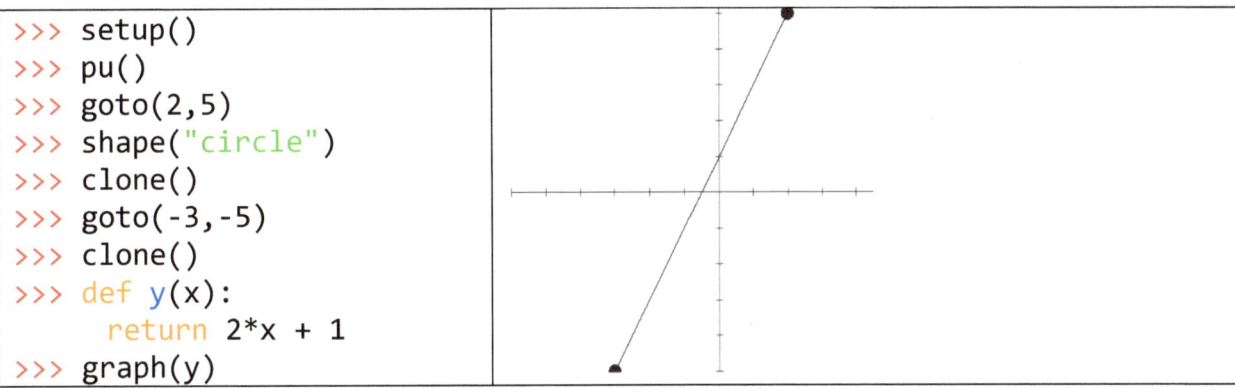

```
>>> setup()
>>> pu()
>>> goto(2,5)
>>> shape("circle")
>>> clone()
>>> goto(-3,-5)
>>> clone()
>>> def y(x):
        return 2*x + 1
>>> graph(y)
```

Yes, it does! Maybe we should just add the graph to our function so we can get instant visual
confirmation:

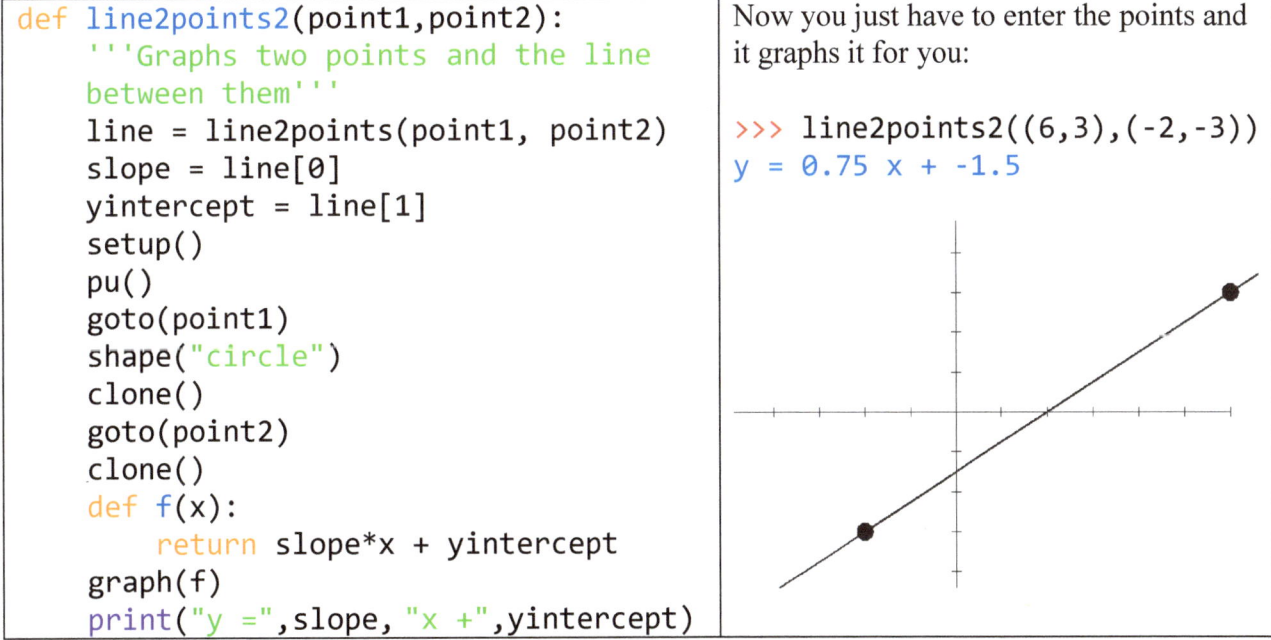

```
def line2points2(point1,point2):
    '''Graphs two points and the line
    between them'''
    line = line2points(point1, point2)
    slope = line[0]
    yintercept = line[1]
    setup()
    pu()
    goto(point1)
    shape("circle")
    clone()
    goto(point2)
    clone()
    def f(x):
        return slope*x + yintercept
    graph(f)
    print("y =",slope, "x +",yintercept)
```

Now you just have to enter the points and
it graphs it for you:

```
>>> line2points2((6,3),(-2,-3))
y = 0.75 x + -1.5
```

Program: Finding the midpoint of two points

The midpoint of two points is simply the midpoint of the x-values and the midpoint of the y-
values. One of the first functions we learned was the average. The first thing I do is import the

"average" function from the arithmetic chapter.

```
from arithmetic import average
def midpt(point1,point2):
    '''Returns the midpoint of two points'''
    return average(point1[0],point2[0]),average(point1[1],point2[1])
```

Here's how to find the midpoint of (6, -1) and (-2, 4).
```
>>> midpt((6,-1),(-2,4))
(2.0, 1.5)
```

Finding intersections of lines

Here's a great example of tools building on other tools. To find the intersection of two lines,

$$y = ax + b \text{ and}$$
$$y = cx + d$$

you set the two right-hand expressions equal to each other and solve for x. Recognize this?

$$ax + b = cx + d$$

We can simply plug those numbers into the "equation" function from the Algebra chapter to solve for x. Plug that into either equation to get the y value. So the user only has to enter the slopes and y-intercepts of two lines to get the intersection point:

```
from algebra import equation
def intersection(a,b,c,d):
    '''Returns the intersection of two lines y = ax + b
    and y = cx + d'''
    x = equation(a,b,c,d) #solve for x
    y = a*x + b  #plug x in to find y
    return (x,y)
```

You just have to have your "equation" function defined in the same module. Here's how to find the intersection of $y = x + 2$ and $y = 5x - 8$:

```
>>> intersection(1,2,5,-8)
(2.5, 4.5)
```

Let's add a grapher:

```
def graphIntersection(slope1, yintercept1,slope2,yintercept2):
```

```
'''Graphs 2 lines and their intersection'''
intpoint = intersection(slope1, yintercept1,slope2,yintercept2)
print(intpoint)
setup()
pu()
goto(intpoint)
shape("circle")
clone()
def f(x):
    return slope1*x + yintercept1
graph(f)
pu()
def g(x):
    return slope2*x + yintercept2
graph(g)
```

Here's how to find and graph the intersection of the lines $$y = \frac{1}{7}x - 5$$ and $$y = -3x + 4$$ The decimal point is necessary because of the fraction in the slope: `>>> graphIntersection(1/7,-5,-3,4)` `(2.86363636363638, -4.5909090909091)`	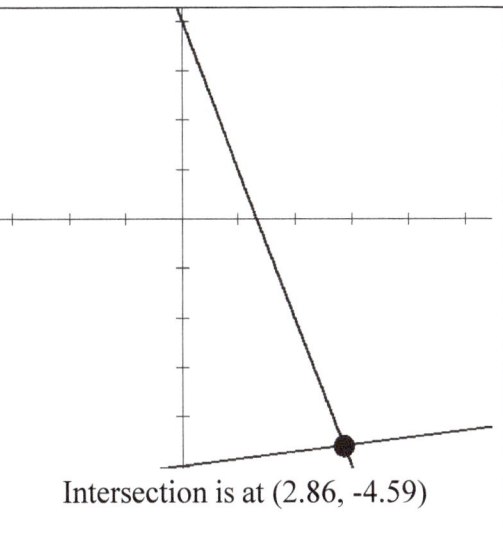 Intersection is at (2.86, -4.59)

Program: Distance between two points

Often you need to know the distance between two points. It's really just the Pythagorean Theorem.

```
from math import sqrt
def distance(point1,point2):
    '''Returns the distance between two points'''
    return sqrt((point1[0] - point2[0])**2 + (point1[1] - \
            point2[1])**2)
```

(The backslash tells Python to continue the next line)

```
>>> distance((5,-3),(-8,12))
19.849433241279208
```

Finding Area of a Triangle Using Heron's Formula

Heron's formula was a work of genius. Given 3 side lengths of a triangle a, b and c, you calculate s, which is half the perimeter:

$$s = \frac{a + b + c}{2}$$

Then the area is found using this formula:

$$A = \sqrt{s(s - a)(s - b)(s - c)}$$

It's not nice to work with if you're doing it by hand, but we could easily code this in Python.

```
def heron(side1,side2,side3):
    '''Returns the area of a triangle given 3 side lengths'''
    semi = (side1 + side2 + side3)/2
    return sqrt(semi*(semi - side1)*(semi - side2)*(semi - side3))
```

To find the area of a triangle with side lengths 4, 5 and 6:

	To check using Geogebra:
```>>> heron(4,5,6)``` ```9.921567416492215```	

Let's build on that formula (and function) by taking the vertex points as input. We just created a function to find distances between points, so we'll put "distance" together with "heron."

```
def heronPoints(point1,point2,point3):
    '''Returns the area of a triangle given 3 side lengths'''
```

```
    side1 = distance(point1,point2)
    side2 = distance(point2,point3)
    side3 = distance(point1,point3)
    return heron(side1,side2,side3)
```

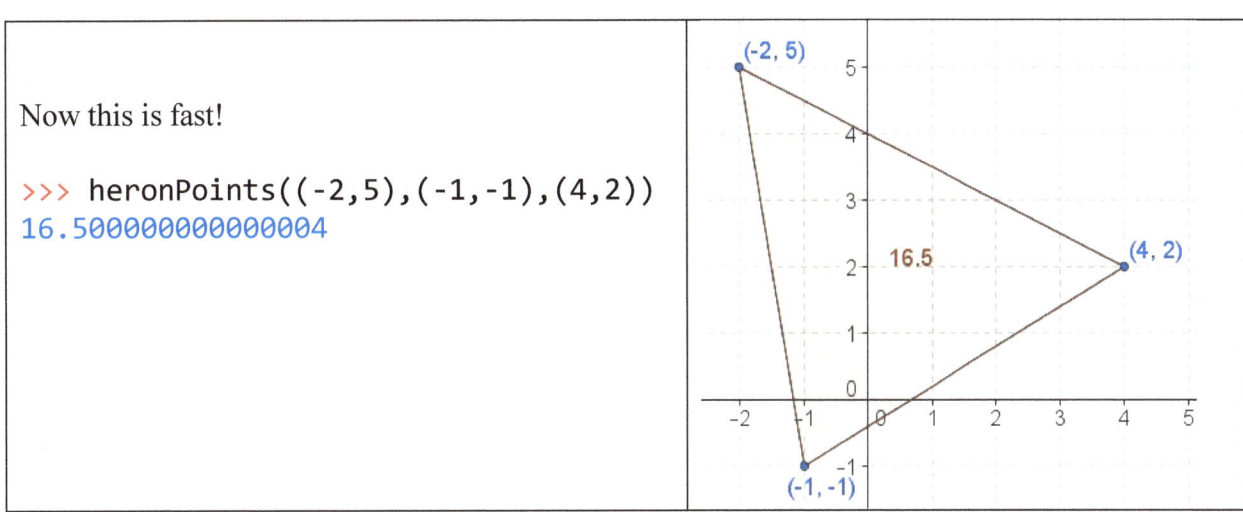

Now this is fast!

```
>>> heronPoints((-2,5),(-1,-1),(4,2))
16.500000000000004
```

Finding Lines Through Points

Program: Line with Given Slope through Given Point

Transform $y - y_0 = m(x - x_0)$ into $y = mx + b$:

```
def line(slope,point):
    '''Returns the slope and y-intercept of a line
    with the given slope through the given point'''
    return slope, point[1]-slope*point[0]
```

Program: Line Perpendicular to Given Line through Given Point

This is an example of a common problem in geometry:

Find the line through the point (2,3) perpendicular to the line with slope $\dfrac{3}{4}$ passing through the point (2,3).

How do you draw a line perpendicular to a given line? You take the slope of the given line, turn it into its reciprocal and change the sign. Here's the Python code to do that:

```
def negativeRecip(number):
    '''Returns the negative reciprocal of a number'''
    return -1/number
```

Now that you know the slope you can use the line function from above to give you the complete equation of the line.

```python
def perpendLine(slope,point):
    '''Returns the slope and y-intercept of a line
    perpendicular to a line with given slope
    through given point'''
    return line(negativeRecip(slope),point)
```

Here's the output.

```
>>> perpendLine(3/4,(2,3))
(-1.3333333333333333, 5.666666666666666)
```

Graph those lines and point to check it. Now enter our lines and point:

```python
>>> def g(x):
        return (3/4)*x

>>> graph(g)
>>> def h(x):
        return -1.33*x + 5.67

>>> graph(h)
>>> pu()
>>> goto(2,3)
>>> shape("circle")
```

The line that passes through the origin is g(x), a generic line with slope ¾. The dot is the point (2,3) and the line through it is h(x), the line perpendicular to g(x).

Perpendicular Bisector

Program: Finding the perpendicular bisector of a segment between 2 points

To find the perpendicular bisector of two points, like (1, 5) and (-2, -3), we're going to put together some functions. First, find the slope of the line between them:

```
>>> line2points((1,5),(-2,-3))
(2.6666666666666665, 2.3333333333333335)
```

So the slope of the line between the two points is 2.67. Next we'll find their midpoint:

```
>>> midpt((1,5),(-2,-3))
(-0.5, 1.0)
```

Now we'll find the line perpendicular to the line with slope 2.67 through the point (-0.5, 1.0). I'll put in a few more digits of the repeating decimal to get more accurate results:

```
>>> perpendLine(2.666667,(-0.5,1))
(-0.3749999531250059, 0.812500023437497)
```

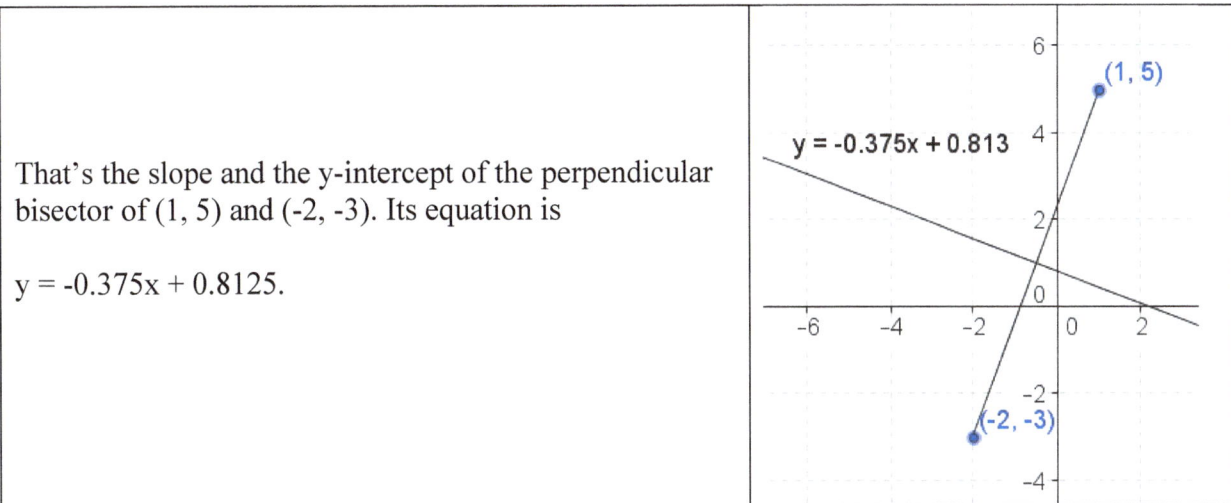

That's the slope and the y-intercept of the perpendicular bisector of (1, 5) and (-2, -3). Its equation is

y = -0.375x + 0.8125.

We can put all those steps together in a perpendicular bisector function:

```
def perpBisect(point1, point2):
    '''Returns the slope and y-intercept of the
    Perpendicular Bisector of 2 points'''
    line = line2points(point1,point2)
    midpoint = midpt(point1,point2)
    return perpendLine(line[0],midpoint)
```

It gives the exact values instantly!

```
>>> perpBisect((1, 5),(-2,-3))
(-0.375, 0.8125)
```

Should we make a grapher to check it?

```
def graphPerpBisect(point1,point2):
    '''graphs two points and the perpendicular bisector
    of the segment between them'''
    setup()
    pu()
    goto(point1)
    shape("circle")
    clone()
    pd()
    goto(point2)
```

```
clone()
pu()
line = perpBisect(point1,point2)
def f(x):
    return line[0]*x + line[1]
graph(f)
```

Here's how to find the perpendicular bisector of the points (1, 5) and (-2, -3):

```
>>> graphPerpBisect((1,5),(-2,-3))
```

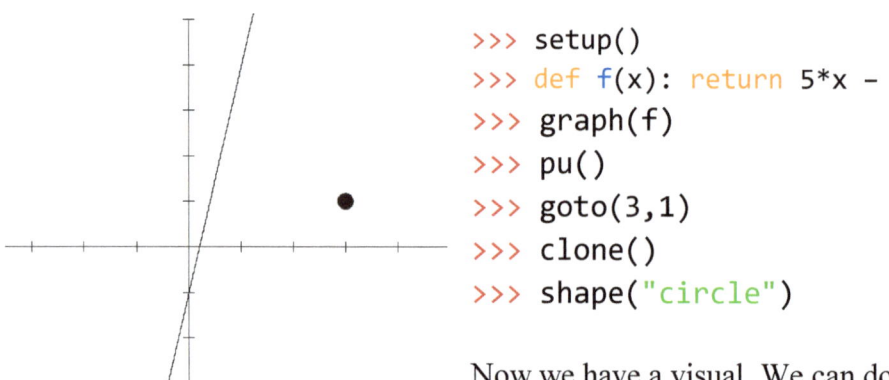

Distance from a point to a line

Here's a task that will require putting together a bunch of programs we have in our file already. Using our grapher, we can easily graph lines and a point. For instance, let's find the distance from the point (3,1) to the line y = 5x − 1. First we'll graph the line and put a dot on the point.

```
>>> setup()
>>> def f(x): return 5*x - 1
>>> graph(f)
>>> pu()
>>> goto(3,1)
>>> clone()
>>> shape("circle")
```

Now we have a visual. We can do the "line through a point perpendicular to a given line" procedure and then find the intersection of the two lines.

```
>>> perpendLine(5,(3,1))
(-0.2, 1.6)
```

That means the line through (3,1), perpendicular to the line with slope 5, is y = -0.2x + 1.6. Now we can intersect those two lines:

```
>>> intersection(5,-1,-0.2,1.6)
```

60

```
(0.5, 1.5)
```

Now we just find the distance from (3,1) to (0.5,1.5):
```
>>> distance((3,1),(0.5,1.5))
2.5495097567963922
```

Can't we just make all that into one function? Then the user could just enter the point and the line and the program would spit out the distance:

```
def distPointLine(point, slope, y-intercept):
    '''Returns the distance from a point to a line
    Line should be entered as slope, y-intercept'''
```

What did we do first? We found the perpendicular line:
```
perp_line = perpendLine(slope,point)
```

Then we found the point of intersection of the two lines:
```
intersect = intersection(slope, y_intercept, perp_line[0],
perp_line[1])
```

Finally we used the distance function to find the distance between those two points:
```
dist = distance(intersect,(point))
return dist
```

Here's the entire code:
```
def distPointLine(point, slope, y_intercept):
    '''Returns the distance from a point to a line.
    Line should be entered as slope, y-intercept'''
    perp_line = perpendLine(slope,point)
    #from the intersection function:
    intersect = intersection(slope, y_intercept, perp_line[0],
                                             perp_line[1])

    dist = distance(intersect,(point))
    return dist
```

Now one line is all it takes to get our result:
```
>>> distPointLine((3,1),5,-1)
2.5495097567963922
```

Program: Point equidistant to 3 points

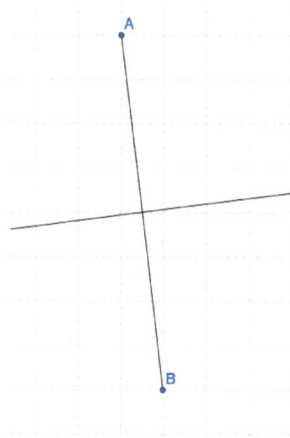

For this exploration we need to think geometrically. Obviously the points equidistant from 2 points make up the perpendicular bisector of the two points.

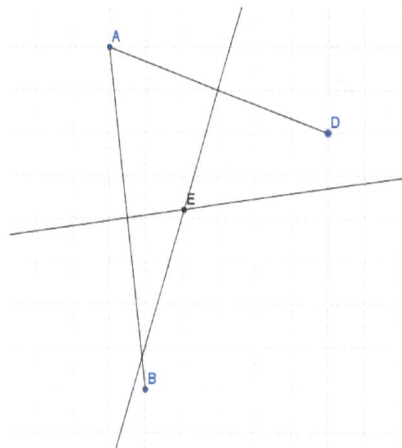

If you put in another point, you simply intersect two perpendicular bisectors!

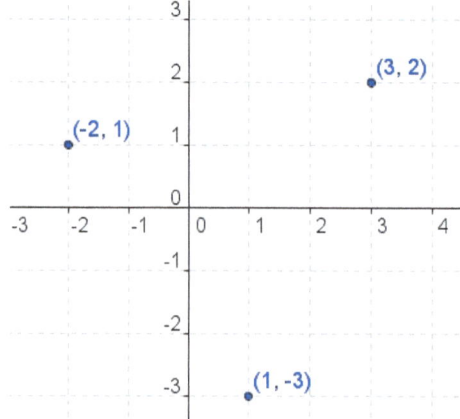

To find the point equidistant to the three points in the figure above (graphed with Geogebra), simply intersect their perpendicular bisectors:

```
def equidPt(point1,point2,point3):
    '''Returns the coordinates of the point
    equidistant to three points'''
    line1 = perpBisect(point1,point2)
    line2 = perpBisect(point2,point3)
    return intersection(line1[0],line1[1],line2[0],line2[1])
```

```
>>> equidPt((-2,1),(3,2),(1,-3))
(0.8043478260869567, -0.02173913043478315)
```

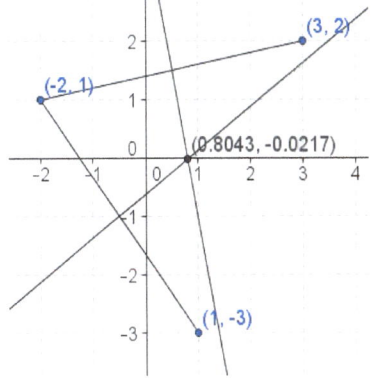

I drew the points in Geogebra, and found the point in the figure above. Using this point, you could find the center and radius of the circle around the three points, called the circumcircle.

Circumcircle

Program: Find Circumcircle

```
def circumcircle(point1,point2,point3):
    '''Returns the center and radius of the circumcircle
    of a triangle given 3 points'''
    center = equidPt(point1,point2,point3)
    radius = distance(center,point1)
    shape("circle")
    pu()
    goto(point1)
    st()
    clone()
    goto(point2)
    clone()
    goto(point3)
```

```
clone()
drawCircle(center,radius)
return center, radius
```

Here are the three points (in Geogebra):

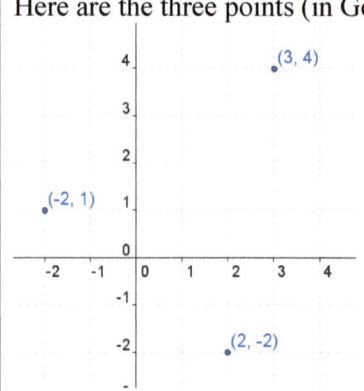

Here's how to find the circle through all three:
```
>>> setup()
>>> circumcircle((3,4),(-2,1),(2,-2))
(1.277777777777778, 1.2037037037037037)
3.28410145388
```

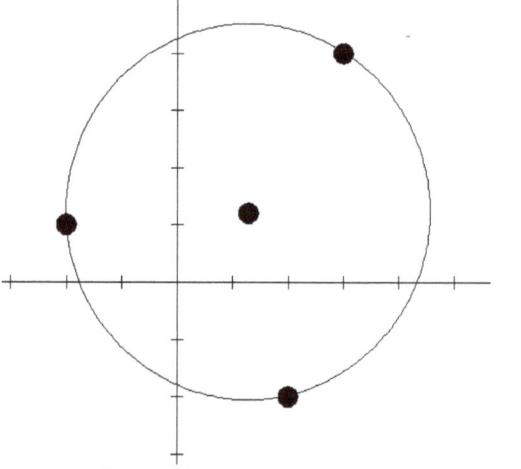

Here's the circle around the points. The `circumcircle` function built on a *lot* of other functions, but now it's just a matter of entering the code on the left.

Centroid

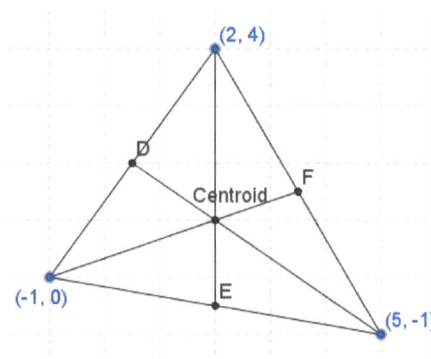

Program: Find the centroid of triangle
The centroid is the intersection of the medians of a triangle. The median is drawn from a vertex to the midpoint of the opposite side. We already have tools for finding the

- midpoint of two points
- line between 2 points
- intersection of two lines

This should be easy!

```
#Returns the centroid of a triangle given 3 points
def centroid(point1,point2,point3):
    median1 = line2points(midpt(point1,point2),point3)
    median2 = line2points(midpt(point3,point2),point1)
    return intersection(median1[0],median1[1],median2[0],median2[1])
```

Here's the centroid of the triangle in the figure:
```
>>> centroid((-1,0),(2,4),(5,-1))
(2.0, 1.0000000000000002)
```
That's (2, 1).

The Geometry Tools

Here's a list of the tools we created in this chapter just by building on previous tools. And I left out a few!

```
drawCircle
line2points
midpt
intersection
graphIntersection
distance
heron
line
perpendLine
perpBisect
distPointLine
equidPt
circumcircle
centroid
```

Here's an extra tool for a textbook favorite: finding the vertex of a parabola.

Program: Finding the vertex of a parabola

```python
def vertex(a,b,c):
    '''Returns the vertex of y = a*x**2 + b*x + c'''
    #This finds the axis of symmetry
    h = -b/(2*a)
    # plug in to find y-value of vertex
    k = a*h**2 + b*h + c
    print("The vertex is (",h, ",",k, ")")
```

So here's how to find the vertex of the parabola $y = 5x^2 - 4x + 9$

```
>>> vertex(5,-4,9)
The vertex is ( 0.4 , 8.2 )
```

Geometry Exercises
(Answers and Solutions on page 141)

1. Point A is (1,5) and Point B is (-3,-1). Find
   a. the distance from A to B.
   b. the midpoint of A and B.
   c. the equation of the line between A and B
   d. the equation of the perpendicular bisector of segment AB

2. What is the distance from the point (4,1) to the line y = 2x + 5?

3. Find the point equidistant to the points (-2, 6), (5,4) and (-4,-1).

4. A triangle is formed by joining the points A(4, 3), B(3, -4) and C(-2, 4).
   a. What is the area of the triangle?
   b. What is the equation of the line through A perpendicular to BC?
   c. What is the center and radius of the circle through A, B and C?

5. Point A is (-1.5, 4.25), point B is (5.25, 5.333), point C is (-4.66, -2.125) and point D is (4.75, -2.25).
   a. Find the equations of lines AB and CD.
   b. What is the intersection point of lines AB and CD?
   c. What is the equation of the line through D perpendicular to line AB?

6. What is the equation of the line with slope 5 through (8,7)?

7. What is the distance from the point (5,-4) to the line y = 2x + 5?

5.   Trigonometry

Sines and Cosines

Trigonometry is like a branch of geometry concerned with right triangles. Sine and cosine are ratios of sides in right triangles. But they became functions of the angles in the triangles, and then their usefulness was extended away from angles altogether! Now they're functions whose output oscillates between -1 and 1 and they're used to model waves, sound, light, heat, electricity and more. Sines and cosines are used to model oscillating behavior in the next exploration.

Harmonographs

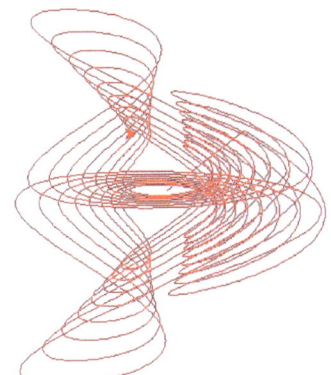

Using the turtle module, you can make harmonograph designs using sine waves and modeling decay with exponential functions.

Sine functions just oscillate back and forth. Create a new file called harmonograph.py and enter this code:

```python
from turtle import *
from math import pi, sin, e

t = 0 #initial time
dt = 0.1 #time increment

clear()
color("red")

for i in range(500):
    speed(0)
    x1 = 100*sin(t)
    setpos(x1,0)
    t += dt
```

Run it and the turtle just goes back and forth. But add this to your loop:
```python
y1 = 100*sin(t + pi/2)
```

and change setpos to `setpos(x1,y1)`

This phase change (pi/2) makes all the difference: a circle!

The original harmonograph devices used pendulums to draw the figures. The figures would change as the pendulums slowed down. This is called "decay" and is modeled using exponential functions. Let's add exponential decay to our x1 and y1 lines:

```
x1 = 100*sin(t)*e**(-t*0.01)
y1 = 100*sin(t + pi/2)*e**(-t*0.01)
```

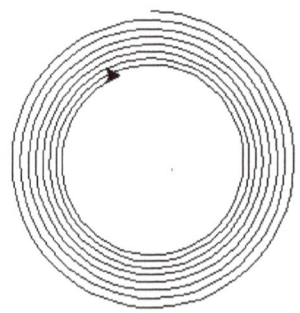

Run it and you get a spiral:

If you change the frequency of the horizontal component of the oscillation, you get a more complicated figure.

Just change t to 2*t like this:
```
x1 = 100*sin(2*t)*e**(-t*0.01)
```

Run it and you'll see this:

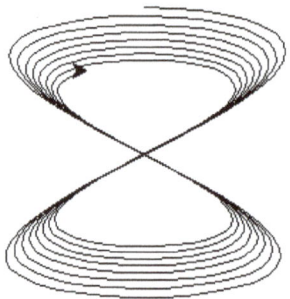

The formula for two pendulums is twice as pretty:
```
x1 = 100*sin(f1*t + p1)*e**(-t*d1) + 100*sin(f2*t + p2)*e**(-t*d2)
```

We'll create a list of frequencies, phase shifts and decay constants. Here's the entire code:
```
from turtle import *
from math import pi, sin, e

#frequencies
f1 = 2
f2 = 6
f3 = 1.002
```

```
f4 = 3

#phase shifts
p1 = pi/16
p2 = 3*pi/16
p3 = 13*pi/16
p4 = pi

#decay constants:
d1 = 0.02
d2 = .0315
d3 = .02
d4 = 0.02

t = 0
dt = 0.1

clear()
color("red")

for i in range(5000):
    speed(0)
    x1 = 100*sin(f1*t + p1)*e**(-t*d1) + 100*sin(f2*t + p2)*e**(-t*d2)
    y1 = 100*sin(f3*t + p3)*e**(-t*d3) + 100*sin(f4*t + p4)*e**(-t*d4)
    setpos(x1,y1)
    t += dt
```

Run it and you'll get a harmonograph!

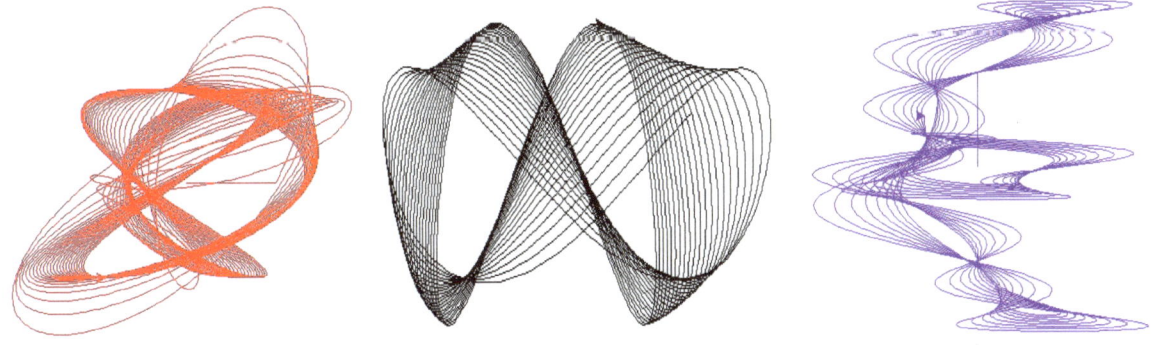

Change the values for frequency, phase shift and decay for different designs. I got the code for these designs on the "Walking Randomly" blog at http://www.walkingrandomly.com/?p=151

Spirograph

Ever play with the Spirograph? Eli Maor has a great chapter on it in his book *Trigonometric Delights*. Here's a figure from his book:

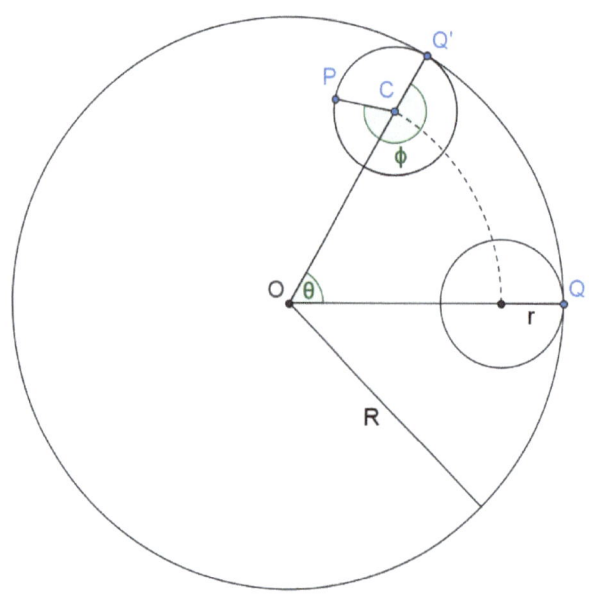

For those familiar with polar equations, it's easy to figure out the coordinates of point C in the figure (the center of the small circle) with respect to O, the origin. But it's more difficult to find out the coordinates of point P with respect to O. This alone would make a great trig exploration for somebody's precalculus class. But Maor points out that the angles of rotation, theta and phi, are not independent; the arc lengths on the two circles must be equal (because their circumferences are connected by the gears). He does some substituting to get everything in terms of theta:

$$x = (R - r)cos\theta + rcos\left[\frac{R - r}{r}\right]\theta$$

$$y = (R - r)sin\theta + rsin\left[\frac{R - r}{r}\right]\theta$$

Now that's a couple of parametric equations that were made for a computer! Let's get a turtle to play Spirograph.

Since Python makes a distinction between uppercase R and lowercase r, we can copy those equations in almost verbatim. But there's no symbol for theta, so just create a variable "theta."

```
#Here are the position formulas for the "pen" point:
    x = (R -r)*cos(theta)+r*cos(((R-r)/r)*theta)
    y = (R -r)*sin(theta)-r*sin(((R-r)/r)*theta)
```

Next we can make a loop so the turtle goes to that x,y coordinate, then increase theta a little, recalculate x and y and go to that coordinate, over and over and over.

```
from turtle import *
```

```
from math import sin, cos

clear()
pu()
theta = 0
R = 300.0 #inner radius of big ring
r = 205.0 #radius of small disk. Change this one
color("red")
pensize(2)
for i in range(500):
    speed(0)
    #Here are the position formulas for the "pen" point:
    x = (R -r)*cos(theta)+r*cos(((R-r)/r)*theta)
    y = (R -r)*sin(theta)-r*sin(((R-r)/r)*theta)
    goto(x,y)#send the turtle to that position
    pd()
    theta += 0.1 #increase theta by a bit
```

But we didn't get what we were expecting.

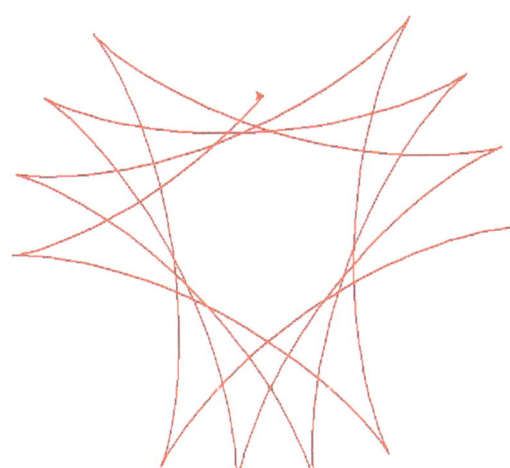

Our design doesn't have nice curved loops, but pointed ends instead. Can you tell what the problem is? Maor's equations assume the "pen" is right on the edge of the smaller circle, not somewhere in the middle.

Let's add a variable p for "proportion" to the second term of the formulas:

```
x = (R-r)*cos(theta)+p*r*cos(((R-r)/r)*theta)
y = (R-r)*sin(theta)-p*r*sin(((R-r)/r)*theta)
```

Now the point can be anywhere on the line. Set p anywhere from 0 to 1. This makes the Spirograph designs we know and love:

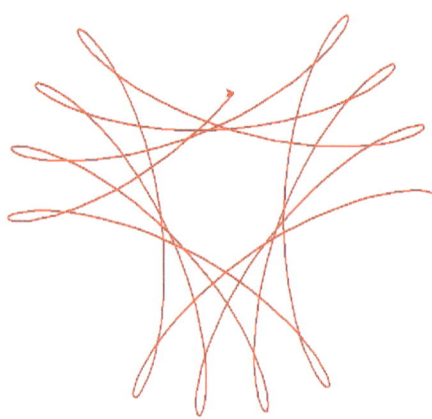

Here's the whole code:

```
#Spirograph via Trigonometric Delights

from turtle import *
from math import sin, cos

clear()
pu()
theta = 0
R = 300.0 #inner radius of big ring
r = 73.0 #radius of small disk. Change this one
p = 0.85 #How close the "hole" is to the center
color("red")
pensize(2)
for i in range(500):
    speed(0)
    #Here are the position formulas for the "pen" point:
    x = (R - r)*cos(theta) + p*r*cos(((R-r)/r)*theta)
    y = (R - r)*sin(theta) - p*r*sin(((R-r)/r)*theta)
    goto(x,y) #send the turtle to that position
    pd()
    theta += 0.1 #increase theta by a bit
```

Play around with r and p to create other designs!

Later in the chapter Maor deals with cases of different ratios of R/r, such as 2, when you simply get a horizontal line. "Even more interesting is the case R/r = 4," when you get a four-pronged thingy:

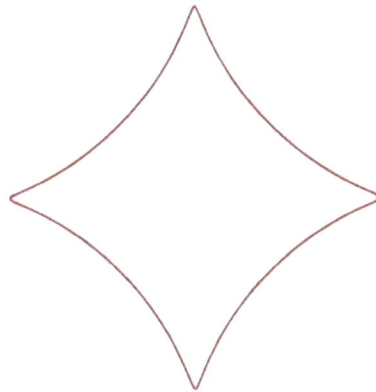

"Interesting"? I don't know, but if you change the proportion variable, you get a slightly different thingy.

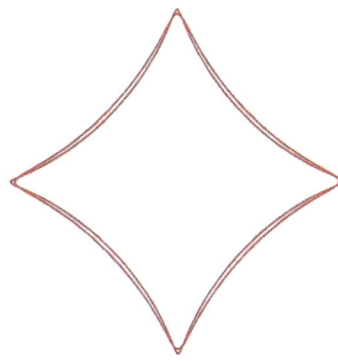

There's another path very near the original one. It's small!

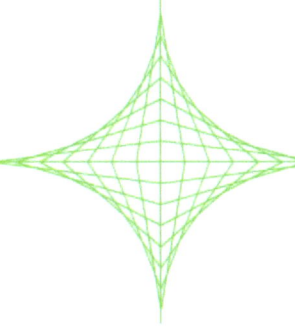

Keep going with this and Maor says you get an "Astroid," which looks like this:

Really? Modify your code to draw a diamond thingy (theta goes from 0 to 2*pi), then lower the proportion variable and draw another one and so on.

Here's what the loop looks like now:

```
for i in range(20): #make this many "thingies"
  while theta <= 6.28: #This makes one 4-pronged thingy
        speed(0)
        #Here are the position formulas for the "pen" point:
        x = (R -r)*cos(theta)+proportion*r*cos(((R-r)/r)*theta)
```

```
        y = (R-r)*sin(theta)-proportion*r*sin(((R-r)/r)*theta)
        goto(x,y) #send the turtle to that position
        pd()
        theta += 0.1 #increase theta by a bit
    proportion -= 0.1 #each thingy gets a different proportion
    theta = 0.0
```

We don't get an astroid but something wholly unexpected:

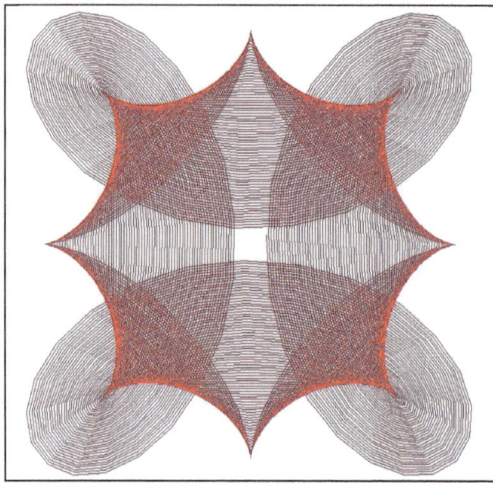

This is something that can't be made with the physical Spirograph, but using computers and extending patterns, a lot of interesting things suddenly become possible!

Let's start calling this figure the Farrell Hypocycloid and see if it catches on.

Even changing the code so the proportion goes up rather than down gets you a cool figure but not an astroid.

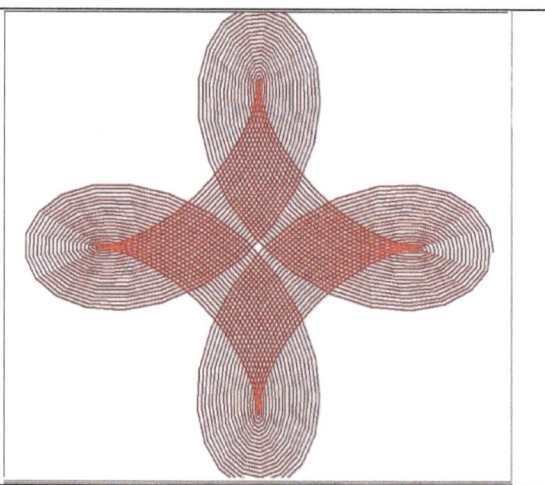

74

Radians

The first change most students notice when beginning "precalculus" is that angles are no longer measured in degrees, but radians. With the rise of calculating machines, there's no longer a good reason to divide the circle up into 360 slices, and radians make finding arc lengths a cinch. But converting between radians and degrees is still a chore, so let's make a computer do it:

Program: Converting Degrees to Radians
You simply want to multiply the degrees by pi/180 and return the radian measure.

```python
from math import pi

def convToRads(degrees):
    '''converts degrees to radians'''
    return degrees*pi/180
```

Check that 90 degrees is pi/2:
```
>>> convToRads(90)
1.5707963267948966
```

Program: Converting Degrees to Radians
And while we're here we might as well create a function to convert radians to degrees. We'll use this function in the future to help create tools for learning Complex Numbers and Calculus.

```python
def convToDegs(radians):
    '''Converts radians to degrees'''
    return radians*180/pi
```

Check that pi/6 is 30 degrees:
```
>>> convToDegs(pi/6)
29.99999999999996
```

Close enough to 30 degrees.

Law of Cosines

A very powerful tool for solving unknown sides or angles in triangles is the Law of Cosines. it's kind of long and complicated:

$$c^2 = a^2 + b^2 - 2abCosC$$

Let's create some Law of Cosines tools.

Program: Solving Triangles using the Law of Cosines

```python
from math import sqrt, acos
def lawofCos_c(a,b,C):
    '''Returns the side opposite the given angle in
    a triangle using the Law of Cosines
    Enter side, side, angle'''
    c = sqrt(a**2 + b**2 - 2*a*b*cos(convToRads(C)))
    return c


def lawofCos_C(a,b,c):
    '''Returns the angle opposite side c in
    a triangle using the Law of Cosines
    Enter 3 sides'''
    C = convToDegs(acos((c**2 - a**2 - b**2)/(-2*a*b)))
    return C
```

To solve for the missing side in a triangle, here's what you enter:
```python
>>> lawofCos_c(7,8,83)
9.96747879549405
```

To solve all three sides and all three angles in a triangle, just use the program below:

```python
def solveTri3sides(a,b,c):
    '''Solves for all three sides and all three angles in a
    triangle using the Law of Cosines
    given 3 sides'''
    C = lawofCos_C(a,b,c) # solve for angle C
    A = lawofCos_C(b,c,a) # solve for angle A
    B = 180 - C - A       # then angle B is easy
    print('a = ',a)       # print the results
    print('b = ',b)
    print('c = ',c)
    print('A = ',A)
    print('B = ',B)
    print('C = ',C)
```

To solve the triangle with sides 11, 12 and 15:
```python
>>> solveTri3sides(11,12,15)
```

76

```
a  =   11
b  =   12
c  =   15
A  =   46.4577809718
B  =   52.2569610468
C  =   81.2852579814
```

Trigonometry Tools

Here's a list of the tools you've created in this chapter:

```
Harmonograph
Spirograph
convToRads
convToDegs
Law Of Cosines
solveTri3sides
```

Trigonometry Exercises
(Solutions on page 142)

1. Convert 50 degrees to radians.

2. How many degrees are in 1 radian?

3. Give all the sidelengths and angles of a triangle where the angle between the 20 cm side and the 17 cm side measures 34 degrees.

4. A triangle contains sides of length 6, 7 and 9 inches. Find all three angles of the triangle.

6. 3D Graphics

Pi3D is an amazing 3-D graphics library. Using Python, you can create art, physics simulations and other dynamic, interactive graphics. Install it on your Pi, Ubuntu or Windows computer by following the directions on their docs at *https://pi3d.github.io/html/index.html*

Programming 3D graphics is a very visual and dynamic way to explore these math topics:

- Coordinate Systems
- Angles
- Distance Formulas
- Translations and Rotations
- Parametric Equations
- Trigonometry

- Matrices
- Velocity and Acceleration
- Vectors
- Projectile Motion
- 3D Shapes

Here's the most basic program, which will create a sphere on the screen:

```python
import pi3d
DISPLAY = pi3d.Display.create()
ball = pi3d.Sphere(z=5.0)
while DISPLAY.loop_running():
  ball.draw()
```

Press F5 or click Run and you'll see what's in Figure 1.

The short program imports the pi3d functions, sets up a display window, creates a Sphere called "ball" and starts an infinite loop where "ball" is drawn. However, there's no way to exit the program. The solution is to add a conditional that checks for the user clicking the ESC button:

```python
import pi3d

#set up the display window
DISPLAY = pi3d.Display.create()

#create a Sphere called 'ball'
ball = pi3d.Sphere(z=5.0)
```

```
#listen for keystrokes
mykeys = pi3d.Keyboard()

#start the display loop
while DISPLAY.loop_running():
    #store keystrokes
    k = mykeys.read()
    if k == 27: #if "ESC" is pressed
        mykeys.close()
        DISPLAY.destroy() #close the display
        break      #end the loop
    ball.draw() #draw the ball
```

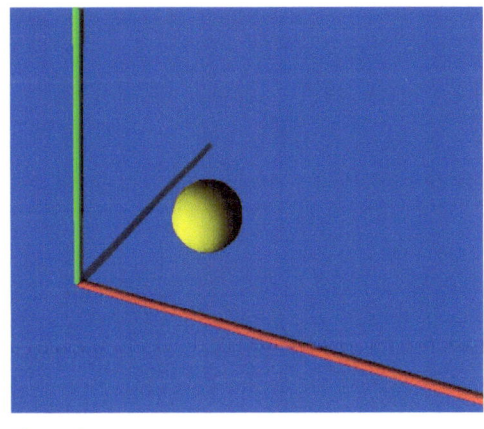

Figure 1

You can change the color of the ball by adding this line after creating the ball:

```
ball.set_material((1.0,1.0,0.0))
```

The "material" can be a color. The numbers are similar to RGB values, where the first number is the amount of **red**, the second number is the amount of **green** and the third number is the amount of **blue**. But instead of going from 0 to 255 it goes from 0 to 1.

Program: Creating x-y-z-Axes

Let's make the x-y-z axes using cylinders. All you need to know to make these cylinders is radius and height. They start off pointing upwards in the y-direction, so we'll rotate the x- and z-axes to point in their proper directions. I made them different colors so we can tell them apart. It's all in this code (after creating the ball):

Figure 2

```
xaxis = pi3d.Cylinder(radius = 0.1,
                      height = 10.0,
                      rz = 90.0) #rotated around z-axis
xaxis.set_material((1.0,0.0,0.0)) # red

yaxis = pi3d.Cylinder(radius = 0.1,
                      height = 5.0)
yaxis.set_material((0.0,1.0,0.0)) # green

zaxis = pi3d.Cylinder(radius = 0.1,
```

```
                    height = 5.0,
                    rx = 90.0) #rotated around x-axis
zaxis.set_material((0.0,0.0,1.0)) # blue
```

Be sure you add this code after "ball.draw()" to draw the axes:

```
xaxis.draw() #draw the axes
yaxis.draw()
zaxis.draw()
```

Here's what it looks like so far:

 All you can see is the ball! We need to "step back" a little. Create a Camera object right after creating the DISPLAY:

```
#set up a camera at (10,10,-15)
CAMERA = pi3d.Camera(eye = (10,10,-15))
```

The camera's "eye" is now at (10,10,-15) but it's still pointing at (0,0,0). Now you can see the axes and the ball.

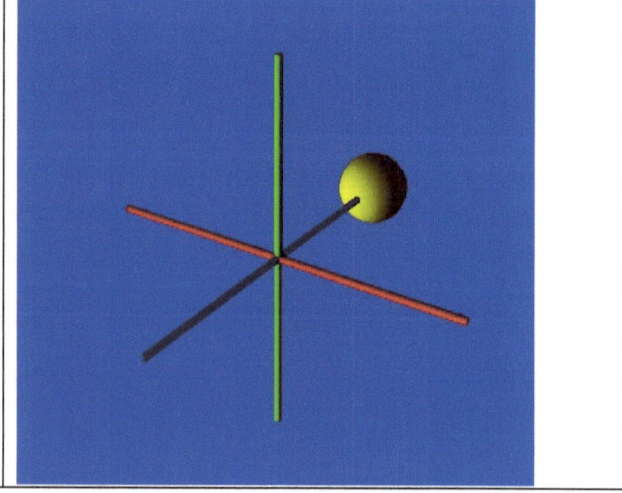

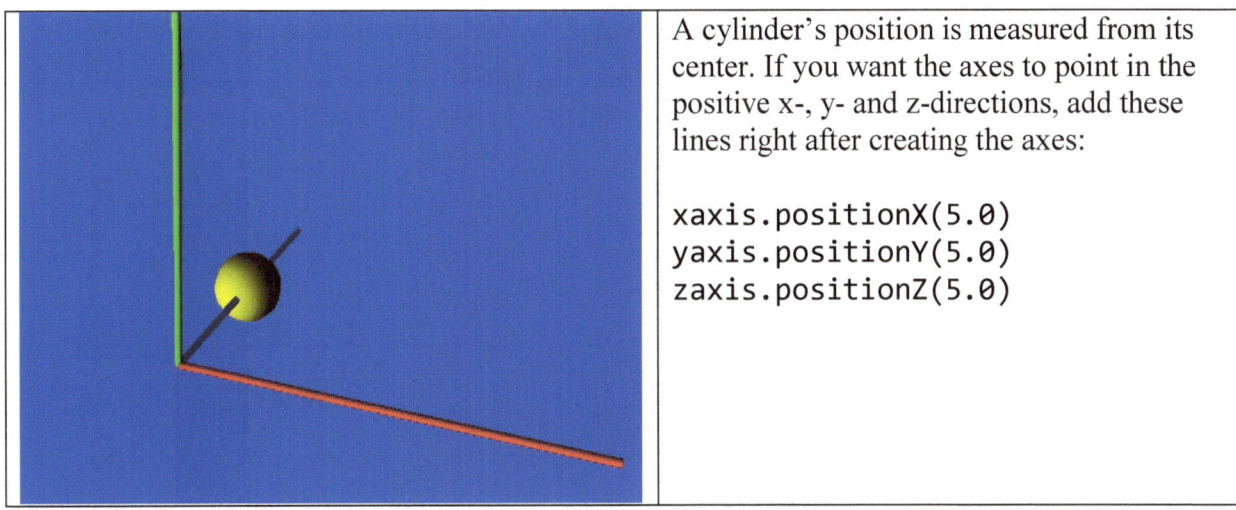

A cylinder's position is measured from its center. If you want the axes to point in the positive x-, y- and z-directions, add these lines right after creating the axes:

```
xaxis.positionX(5.0)
yaxis.positionY(5.0)
zaxis.positionZ(5.0)
```

Absolute location using `position()`

You can change the position of objects by making the position a function of a variable such as t for time. Here's how to use sines and cosines to make the ball move in a circular path. You have to import those functions from Python's numpy module (which you have to install) by adding this to the beginning of the code:

```
import pi3d
from numpy import sin, cos
```

Change the code to:

```
t = 0 #starting t
dt = 0.1 #time steps
#start the display loop
while DISPLAY.loop_running():
    ball.position(5*sin(t),
    0,
    5*cos(t))
    t += dt #increment time
```

Now you can see the ball traveling in a circular path.

Vectors

Vectors are extremely useful tools in mathematics and in Python we can either use lists or numpy arrays. If we wanted to move our yellow ball down 1 unit, we could just add a vector and have it translate by that vector before we draw it. Python's numpy module has an array function that does this really fast, so add "array" to the list of numpy functions you import:

```
from numpy import sin, cos, array
```

Then you can create a 3D array for translating in the x, y, and z directions in one line of code
After creating the ball, add this:

```
#update ball's position
ballv = array((0,-1,0))
ball.translate(*ballv)
```

Run it and the ball will move 1 unit downward. It happened so fast we didn't see it. We could put that line of code in the while loop and make the ball keep moving downwards:

```
#start the display loop
while DISPLAY.loop_running():
    #update ball's position
    ballv = array((0,-1,0))
    ball.translate(*ballv)
```

That makes the ball's y-value decrease forever. Run it and you'll see it just keeps flying downwards out of sight. If we put in a conditional before ball.draw(), it will stop the ball when its y-value gets to -10:

```
    #update ball's position
    ballv = array((0,-1,0))
    if ball.y() <= -10: #if the ball gets too far down
        ballv = array((0,0,0))   #stop the ball
    ball.translate(*ballv)
```

When you run this, you can see the ball stops when its y-value reaches -10.	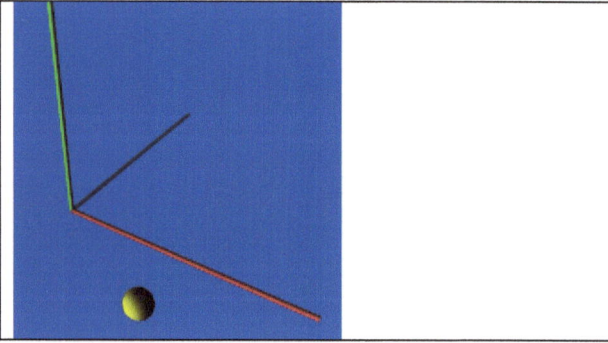

Falling Objects

Using Pi3D and vectors you can answer questions on how long it takes an object to reach the ground, like this one:

If an object is dropped from a height of 100 meters, how long will it take to reach the ground? (Neglect air resistance.)

We'll use the code from the above example and change the initial position:

```
ball = pi3d.Sphere(y=100.0)
```

Change the length of the y-axis:

```
yaxis = pi3d.Cylinder(radius = 0.1,
                      height = 100.0)
```

The only force operating on an object is gravity. We'll create a vector for the force pulling 9.8 meters in the negative y-direction:

```
fgrav = array((0,-9.8,0))
```

We'll also create a vector for the velocity of the ball. Its initial velocity is 0.
```
ballv = array((0.0,0.0,0.0)) #initial velocity of the ball
```

We'll be running a loop and we want to know how long it takes for the ball to hit the ground. That means we need a variable for time, and a variable for the increment of time.

```
t = 0       #time parameter
dt = 0.1    #change in time
```

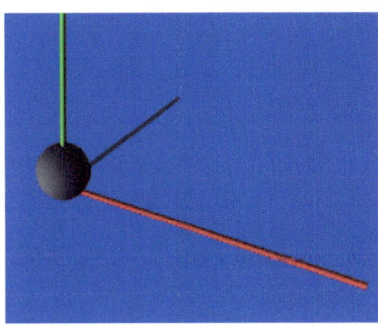

Here's what to put in the while loop. First we'll update the ball's velocity by the force of gravity. Notice everything is multiplied by dt since we're dealing with a fraction of a second here. If the ball's velocity is -5 meters per second, we have to realize it's only going to move the ball a tenth of that distance in a tenth of a second.

Figure 3

```
            ballv += fgrav*dt #update ball's velocity by
force of gravity
ball.translate(*ballv*dt)   #translate ball by its velocity
```

We only need the loop to continue until the ball's y-coordinate is 0. Then it prints t and closes the display.

```
if ball.y() <= 0: #if ball reaches the ground
    print('t = ', t) #print out the time it took
```

```
    mykeys.close()
    DISPLAY.destroy() #close the display
    break       #end the loop
```

```
t += dt #make time go up a step
```

But if we run it we'll only see the bottom of the drop, like in Figure 3

Let's change the Camera code a little so we'll see the whole scene:

```
CAMERA = pi3d.Camera(at = (0.0,50.0,0.0),
                     eye = (0,0,-150.0))
```

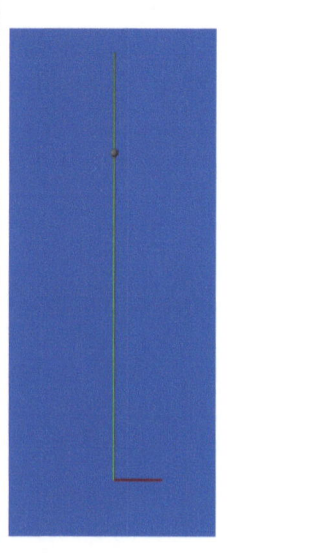

And when the ball makes it to the ground, it prints out the time it took in the terminal: 4.51 seconds.

At least it printed this out in Ubuntu and Windows. If you run this file using the terminal, it will print out.

```
t = 4.51
Press any key to continue . . .
```

Here's the entire code:

```
import pi3d
from numpy import array

#set up the display window
DISPLAY = pi3d.Display.create()

ball = pi3d.Sphere(y=100.0)
```

84

```python
#set up camera object
CAMERA = pi3d.Camera(at = (0.0,50.0,0.0),
                     eye = (0,0,-150.0))


#create axes
xaxis = pi3d.Cylinder(radius = 0.1,
                      height = 10.0,
                      rz = 90.0)
xaxis.set_material((1.0,0.0,0.0)) #x-axis is red


yaxis = pi3d.Cylinder(radius = 0.1,
                      height = 100.0)
yaxis.set_material((0.0,1.0,0.0)) #y-axis is
green


zaxis = pi3d.Cylinder(radius = 0.1,
                      height = 10.0,
                      rx = 90.0)
zaxis.set_material((0.0,0.0,1.0)) #blue
xaxis.positionX(5.0)
yaxis.positionY(50.0)
zaxis.positionZ(5.0)


t = 0 #time parameter
dt = 0.01 #change in time
ballv = array((0.0,0.0,0.0))      #ball's initial velocity
fgrav = array((0,-9.8,0))         #force due to gravity


#listen for keystrokes
mykeys = pi3d.Keyboard()


#start the display loop
while DISPLAY.loop_running():
    #store keystrokes
    k = mykeys.read()
    if k == 27: #if "ESC" is pressed
        mykeys.close()
        DISPLAY.destroy() #close the display
        break      #end the loop
```

Figure 4

```
#update ball's position
ballv += fgrav*dt #update ball's velocity by force of gravity
ball.translate(*ballv*dt) # translate ball by its velocity
if ball.y() <= 0: #if ball reaches the ground
        print('t = ', t) #print out the time it took
        mykeys.close()
        DISPLAY.destroy() #close the display
        break       #end the loop
t += dt
ball.draw() #draw the ball
#draw the axes
xaxis.draw()
yaxis.draw()
zaxis.draw()
```

The Solar System Model

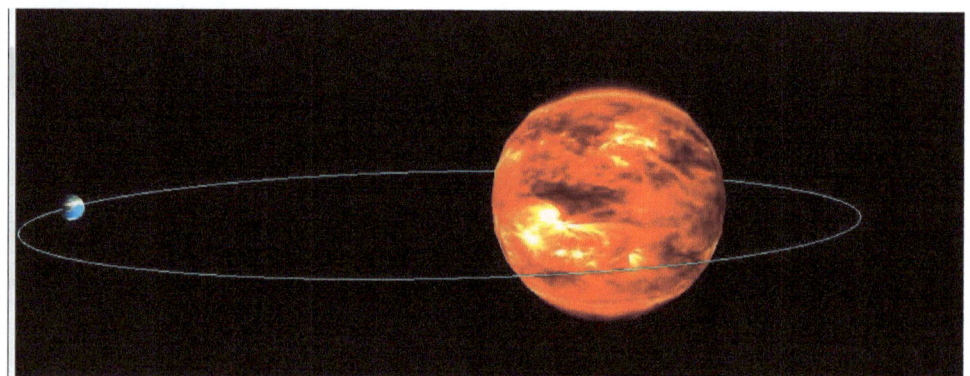

Copernicus said the planets move in perfectly circular orbits. Kepler said the orbits are elliptical. How would you make a model of this to figure it out? What do you do if the direction and magnitude of a vector is always changing? For example, if you're making a solar system model the velocity of the orbiting planet and the force of attraction between the two bodies keeps changing.

$$\vec{F} = -\frac{GmM}{r^2}\hat{r}$$

$\vec{F}$ is the vector of the force of attraction between the Sun and Earth.
G is the Gravitational Constant. It's really small.
m is the mass of the earth. It's really big.
M is the mass of the Sun. It's really, really big.
r is the distance between the two bodies
$\hat{r}$ is the Unit Radial Vector, the one-unit long vector going from the Sun to the Earth.

The negative sign means the force of gravity is going in the opposite direction than the Unit Radial Vector: from the Earth towards the Sun.

Let's create some spheres to be our Sun and Earth. We'll also create some vectors to represent the Earth's motion and the gravitational force pulling it into the Sun.

```python
import pi3d
from numpy import sqrt, array

#create display
DISPLAY = pi3d.Display.create(x=100,y=100)
```

We want the background of our solar system to be black.
```python
DISPLAY.set_background(0.0,0.0,0.0,1) #sets color black
```

We'll set up our camera to be far away from the center:
```python
CAMERA = pi3d.Camera(eye = ((0.0,
0.0, -500)))
```

Now we can create the spheres for our sun and earth.
```python
#Create our objects:
sun = pi3d.Sphere(radius = 100.0)
```

Figure 5

```python
#default position is (0,0,0)
earth = pi3d.Sphere(radius=10, x=200)
```

Now we can add the code for listening for keystrokes to close the display.

```python
#listen for keystrokes
mykeys = pi3d.Keyboard()

#start the display loop
while DISPLAY.loop_running():
    #store keystrokes
    k = mykeys.read()
    if k == 27: #if "ESC" is pressed
        mykeys.close()
        DISPLAY.destroy() #close the display
        break      #end the loop
```

Finally we draw our objects inside the while loop:

```python
sun.draw()
earth.draw()
```

Our scene now looks like Figure 5.

You can use the actual values from an astronomical data chart if you like, but I just want it to look realistic. Speaking of realistic, you can just choose a color for the Earth (blue? green?) but you can download an "earth" picture and wrap it around the sphere.

Here's one I found when I did a web search for "earth texture." I saved it in the same place I have my Python file and called it "worldmap.gif" since it happens to be a gif file.

Before creating the planets, put in the code for shader and texture:

```python
shader = pi3d.Shader('uv_light')
earthimg = pi3d.Texture('worldmap.gif')
```

	Change your earth.draw() line to this and you should get a more realistic earth: `earth.draw(shader, [earthimg])`

Download a sun image as well and do the same thing. They both use "shader."

```
sunimg = pi3d.Texture('sun.jpg')
```

and add this to "sun.draw()":
```
sun.draw(shader,[sunimg])
```

Here's how it looks:

The sun looks a little dark. And where is the light coming from? The sun should be the source of the light! Pi3d can handle that. Define "sunlight" before creating the sun Sphere and add it to the sun code:

```
sunlight = pi3d.Light(lightamb=(1.0, 1.0, 1.0))
sun = pi3d.Sphere(light = sunlight, radius = 100.0)
```

	And create an earthlight too:
	`earthlt = pi3d.Light(is_point=True,`
	`lightpos=(0,0,0),`
	`            lightcol = (500000,`
	`            500000,500000),`
	`            lightamb = (0.05,`
	`            0.05,0.05))`
	
	`earth = pi3d.Sphere(light=earthlt,`
	`            radius=10,x=200.0)`

Much better! Now that they look fairly realistic, let's get the earth to move. First we have to give it a velocity. That'll be an array of three numbers, the velocity in the x, y and z directions:

```
earthv = array([0.0,-1.0,-8.2])
```

You can play around with those values and see what happens when you run it. Then the force of gravity will also be an array. But it's based on Newton's complicated formula above. First we set up a few lines which will give us the important earth-sun distance:

```
xyz = array([earth.x(), earth.y(), earth.z()]) #radial vector
dist = (xyz * xyz).sum()**0.5 #earth-sun distance
```

The vector from the sun to the earth is the radial vector (along the radius of a circle), and if you divide it by its distance, you get a 1-unit long vector pointing from the sun towards the earth.

```
unitRadialVector = xyz / dist
```

I replaced "G*m*M" in Newton's formula with 10,000 because they're all constants. You can experiment with using the real values (get ready to review scientific notation!) but this works, too:

```
Fgrav = -10000.0 * unitRadialVector / dist**2   #force of gravity
```

Now we update the earth's velocity and translate it the same way we did for the falling object.

```
earthv += Fgrav   #update earth's velocity by gravity
earth.translate(*earthv) # update earth's position by its velocity
```

Run this and you'll see the earth revolving around the sun! Here's the whole code, with a mouse object to control how the camera relocates so you can rotate around the scene.

```
import pi3d
from numpy import array, sqrt

#set up display window
DISPLAY = pi3d.Display.create(x=100,y=100)
DISPLAY.set_background(0.0,0.0,0.0,1)

#create shader and textures
shader = pi3d.Shader('uv_light')
sunimg = pi3d.Texture('sun.jpg')
earthimg = pi3d.Texture('worldmap.gif')

#create camera object
CAM = pi3d.Camera(eye=(0.0,0.0,-500))

#create mouse object to rotate scene
mymouse = pi3d.Mouse(restrict=False)
mymouse.start()

#set up the lights
sunlight = pi3d.Light(lightamb = [1.0,1.0,1.0])
earthlt = pi3d.Light(is_point=True, lightpos=(0,0,0),
                    lightcol = (500000,500000,500000),
```

```python
                      lightamb = (0.05,0.05,0.05))

#create our objects
sun = pi3d.Sphere(light=sunlight,
                      radius=100.0,
                      sides = 24)
earth = pi3d.Sphere(light=earthlt,radius=10,x=200.0)

#velocity vector for earth
earthv = array([0.0,-1.0,-8.2])

#listen for keystrokes
mykeys = pi3d.Keyboard()

#start the display loop
while DISPLAY.loop_running():
    k = mykeys.read()
    if k == 27: #if ESC key is pressed
        mykeys.close()
        DISPLAY.destroy()
        break
    #radialVector:
    xyz = array([earth.x(),earth.y(),earth.z()])
    #sun-earth distance
    dist = (xyz * xyz).sum()**0.5

    #unit radial vector
    unitRadialVector = xyz / dist

    #create gravity vector:
    fgrav = -10000.0 * unitRadialVector / dist**2

    earthv += fgrav #update earth's velocity by gravity
    earth.translate(*earthv) #change earth's position by velocity
    earth.rotateIncY(2.1) #earth spins!

    u,v = mymouse.position() #x- and y-coordinates of mouse position
    #rotate the camera around the scene
    CAM.relocate(-u*0.5,v*0.5, #horizontal and vertical rotation
            [0.0,0.0,0.0],   #where camera is pointing
            [-200.0,-200.0,-200.0]) #distance of camera
    sun.draw(shader,[sunimg])
    earth.draw(shader,[earthimg])
```

This has obviously been a very quick introduction to programming 3D graphics. Pi3D has many more possibilities that could take up a whole book!

7. Recursion and Fractals

Recursion is an interesting phenomenon where the computer can execute a function that's defined within itself! An example will help explain:

Program: print using recursion

```python
def print_down(num):
    print(num)
    print_down(num - 1)
```

If you run the program and try to execute the code, like

```python
>>> print_down(num)
```

you'll see the computer print numbers from 10 on down to around -1000 and then Python will stop. If you want to make it stop at 0, for example, you'll need a conditional:

```python
def print_down(num):
    if num > 0:
        print(num)
    else:
        print("Blastoff! ")
        return
    print_down(num - 1)
```

The "return" line makes it exit the function after printing "Blastoff!" Then you'll get this:

```python
>>> print_down(10)
10
9
8
7
6
5
4
3
2
1
Blastoff!
```

Program: Factorials

The factorial operation is defined recursively:

3! = 3 * 2 * 1
4! = 4 * 3 * 2 * 1
5! = 5 * 4 * 3 * 2 * 1

You can write a recursive program to return the factorial of a number n:

```
def factorial(n):
    if n == 0: return 1
    return n * factorial(n - 1)
```

```
>>> factorial(7)
5040
```

To check:
```
>>> 2*3*4*5*6*7
5040
```

Recursion is most striking when you see it in graphics, even using turtles. All you have to do is return the turtle to its starting point.

Fractal Trees

```
from turtle import *
def y(): #How to draw a Y
    fd(100)
    rt(30)
    fd(70)
    bk(70)
    lt(90)
    fd(50)
    bk(50)
    rt(60)
    bk(100)
```

If you set the heading to 90 degrees (straight up), the y will look like this:

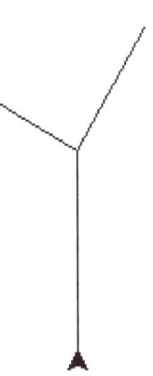

Here's how you can get as many y's as you want. It's an interesting trick. You make the turtle go forward and make a y, turn, make another y and then go back to its starting point.

```
def y(distance, level):
    speed(0)
    if level > 0:
        fd(distance) #go up the trunk
        rt(30)
        y(distance*.8,level-1) #left branch
        lt(90)
        y(distance*.7,level-1) #right branch
        rt(60)
        bk(distance) #back down the trunk
```

Now executing
```
>>> setheading(90)
>>> y(100,5)
```
will get you a tree:

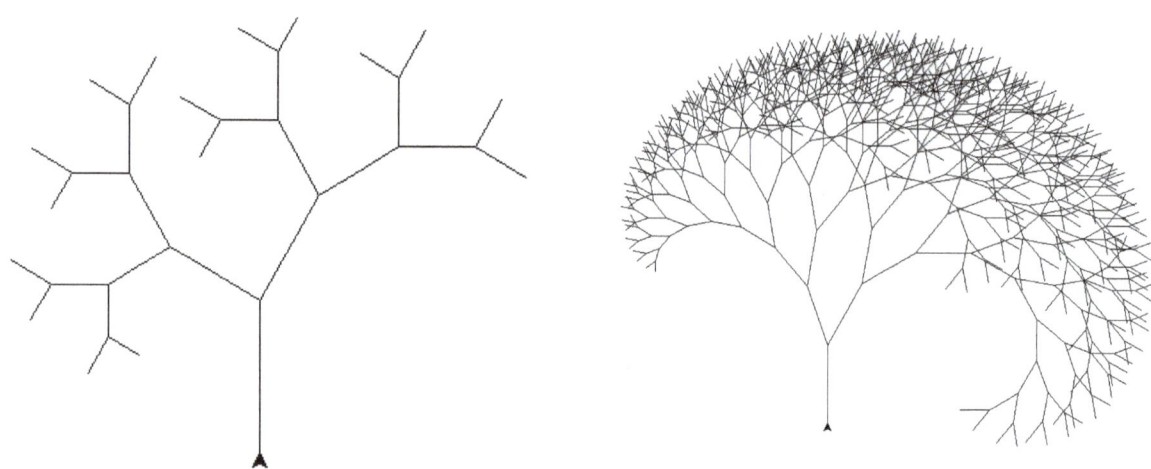

Play around with the lengths, the angles and the levels to see what different kind of trees you can generate!

Pythagorean Tree

If you can create a square function, and make smaller squares at the upper corners of the square, you can make a "Pythagorean Tree" like this one:

First you start off by creating a square function:

```python
def square(length):
    '''square starts from bottom left corner.'''
    color('purple')
    begin_fill() #fills the shape with color
    for i in range(4):
        fd(length)
        rt(90)
    end_fill()
```

After that, you use the square function inside the tree function:

```python
def tree(length,level):
    speed(0)
    if level == 0: return
    pu()
    fd(length/2)
    pd()
    lt(90)
    fd(length/2)
    rt(90)
    square(length) # the big square
    fd(length)
    rt(90)
    fd(length/2)
    lt(135)
    tree(length/(2**0.5),level-1) # the left square
    rt(90)
    tree(length/(2**0.5),level-1) # the right square
```

```
    pd()
    rt(45)
    fd(length/2)
    rt(90)
    fd(length)
    rt(90)
    fd(length/2)
    rt(90)
    pu()
    bk(length/2)

setpos(0,-300)
clear()
seth(90) #face upwards
tree(150,4)
```

Execute it, using different lengths and levels!

Program: Pythagorean Tree with angle offset
If you know your sines and cosines, you can draw a tree with any angle you want between the squares.

```
def pytree2(length, angle,level):
    angledeg = angle*pi/180 #convert angle to radians
    if level == 0: return
    square(length)
    fd(length)
    lt(angle)
    pytree2(length*cos(angledeg),angle,level-1)
    rt(90+angle)
    fd(length)
    lt(90+angle)
    pu()
    fd(length*sin(angledeg))
    rt(90)
    pd()
    pytree2(length*sin(angledeg),angle,level-1)
    rt(90)
    pu()
    fd(length*sin(angledeg))
```

```
pd()
rt(angle)
for i in range(2):
    fd(length)
    rt(90)
```

Koch Snowflake

This is a famous fractal:

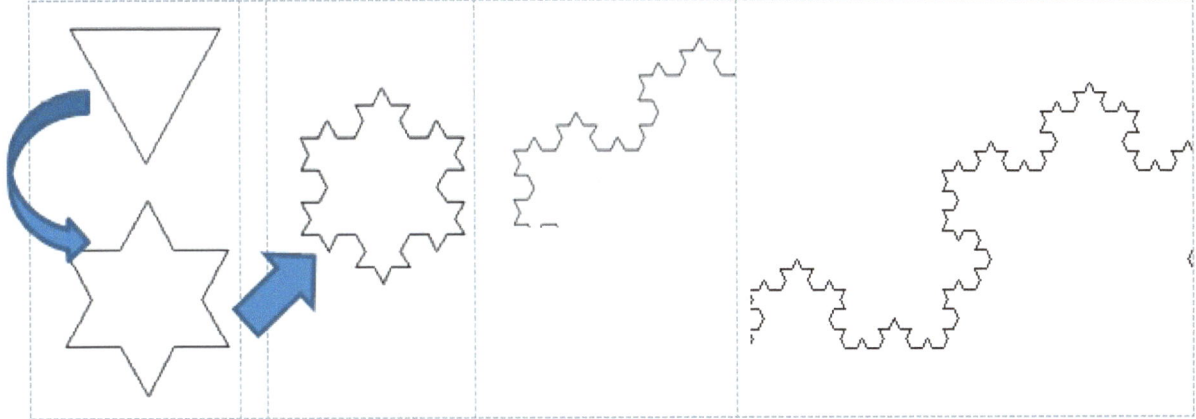

You start with a triangle and add an "angle" to every segment at each step.

So let's start with the code to draw a triangle:

```
def snowflake(length):
    speed(0) #fastest speed
    for i in range(3):
        fd(length)
        rt(120)
```

Now how do we replace every segment with an angle or "bump"?

We have to stop a **third** of the way in,	
turn left, make a segment,	
turn right, make a segment,	
turn left again and make the final segment. All these segments are the same length.	

This function should draw a side:

```
def side(length):
    fd(length/3)
```

```
lt(60)
fd(length/3)
rt(120)
fd(length/3)
lt(60)
fd(length/3)
```

Here comes the interesting part. Add the level and change "fd" to "side"!

```
def side(length,level):
    if level == 0:
        fd(length)
        return
    side(length/3,level - 1)
    lt(60)
    side(length/3,level - 1)
    rt(120)
    side(length/3,level - 1)
    lt(60)
    side(length/3,level - 1)
```

And the snowflake function should include the side function:

```
def snowflake(length, level):
    speed(0)
    for i in range(3):
        side(length, level)
        rt(120)
```

Now execute the snowflake function for different lengths and levels. For example:

```
color('red')
snowflake(200,4)
ht() #hide the turtle
```

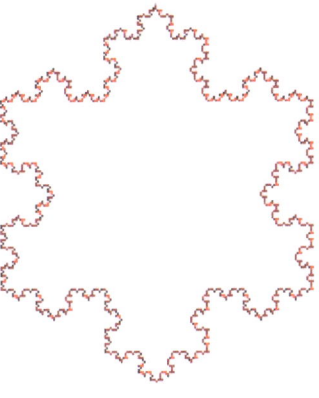

Sierpinski Triangle

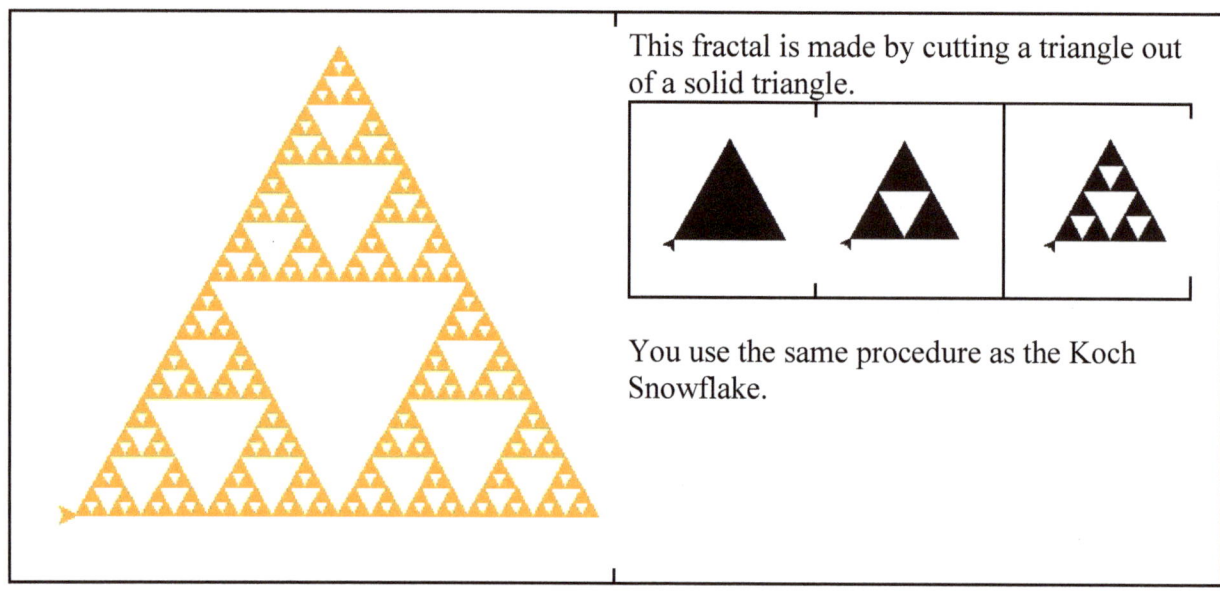

This fractal is made by cutting a triangle out of a solid triangle.

You use the same procedure as the Koch Snowflake.

Of course you know the code for making a triangle:

```python
def sierpinski(side_length):
    speed(0) #fastest speed
    for i in range(3):
        fd(side_length)
        rt(120)
```

But this fractal is only a triangle if the level is zero. If the level is higher, it makes another triangle at each of the corners.

```python
def sierpinski(side_length, level):
    speed(0)
    if level == 0: return
    for i in range(3):
        sierpinski(side_length/2,level - 1)
        fd(side_length)
        rt(120)
```

When you enter this in the shell:
```python
>>> sierpinski(100,2)
```

You get a "level 2" Sierpinski triangle:

100

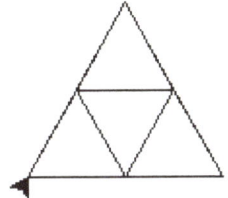

If you'd like the triangles filled in, add the "fill" lines to your code:

```python
def sierpinski(side_length, level):
    speed(0)
    if level == 0: return
    begin_fill()

    for i in range(3):
        sierpinski(side_length/2,level - 1)
        fd(side_length)
        rt(120)
    end_fill()
```

And you start to get the famous fractal.

8. Matrices

Matrices are used extensively in graphics for transformations. Don "The Mathman" Cohen wrote a book called *Changing Shapes With Matrices* which I highly recommend. In this chapter we'll take a look at how to transform shapes with matrices using Python.

Entering Matrices as Lists

A matrix is really just a list, and if it has more than one row, each row is its own list. For example, the matrix

$$A = \begin{bmatrix} 1 & 2 \\ -3 & 4 \end{bmatrix}$$

can be entered into Python like this:

```
A = [[1,2],[-3,4]]
```

or to make it look more like a matrix, you can press ENTER after the commas:

```
A = [[1,2],
     [-3,4]]
```

Here's how to enter a 3x3 matrix as a list of lists in Python: $$B = \begin{bmatrix} 1 & 2 & -3 \\ 4 & -6 & 0 \\ 7 & 1 & -9 \end{bmatrix}$$	`B = [[1,2,-3],[4,-6,0],[7,1,-9]]` Or press ENTER after a comma: ```B = [[1,2,-3],` ` [4,-6,0],` ` [7,1,-9]]```

Transforming Points using Matrices

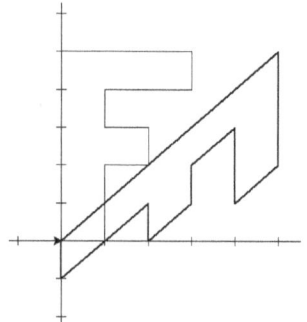

The Python turtle will help us draw figures and their transformations. We'll import the turtle module and our "setup" function to draw a grid:

```
#Grapher with tick marks at whole numbers
from turtle import *
from algebra import setup
```

Now we'll find out the points we need to make an "F":

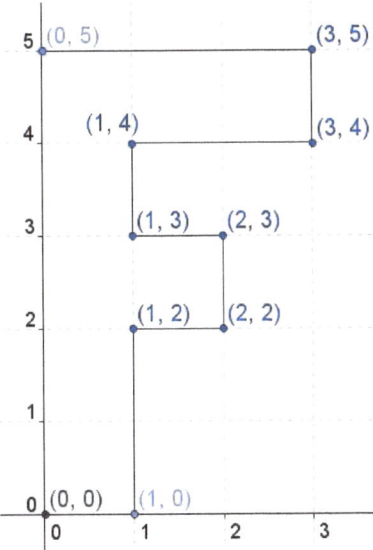

So our points are (0, 0), (1, 0), (1, 2), (2, 2), (2, 3), (1, 3), (1, 4), (3, 4), (3,5), (0, 5) and (0, 0) again. The turtle is going to draw the F so it'll return to its starting point.

Remember in the Geometry chapter we expressed a point as a list, so this list of points will be a list of lists, or a matrix.

```
#The points of the F
fmatrix = [[0,0],
           [1,0],
           [1,2],
           [2,2],
           [2,3],
           [1,3],
           [1,4],
           [3,4],
           [3,5],
           [0,5],
           [0,0]]
```

This next function will draw the F.

```
def drawf(matA):
    pu()
    goto(0,0)
    pd()
    for point in matA: #the turtle goes to all the points in the list
```

```
goto(point[0],point[1])
```

Multiplying Matrices

There's a formula to multiplying **each row** of our 11x2 matrix by a 2x2 matrix:

$$[a \quad b]\begin{bmatrix} c & d \\ e & f \end{bmatrix} = [ac + be \quad ad + bf]$$

This next function will multiply any matrix (a) with 2 columns by a 2x2 matrix (b) and return the product matrix. What you're doing is creating an empty list called "newmatrix" and when you do the row multiplication you'll append the product to the list

```
def multMatrix(a,b):
    '''Multiplies two matrices.
    b is a 2x2 matrix'''
    newmatrix = []
    for i in range(len(a)): # repeat for every row in matrix a
        newmatrix.append([])
        for j in range(2): #because b only has two columns
            newmatrix[i].append(a[i][0]*b[0][j]+a[i][1]*b[1][j])
    return newmatrix
```

Matrix b will be the transformation matrix, for rotating or scaling or reflecting the points in the first matrix. We'll start off with a simple one from Don Cohen's book.

```
#Define Transformation matrix:
transmatrix = [[0,-1],
               [1,1]]

#Multiply F-matrix by transformation matrix
newmat = multMatrix(fmatrix,transmatrix)

setup()
drawf(fmatrix)
pensize(2) #makes transformed F thicker
drawf(newmat)
```

So the output will look like the first figure in the chapter:

104

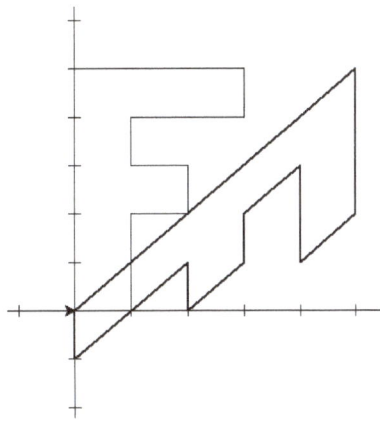

Check out the x- and y-values of the beginning points and transformed points and see if you can see a pattern. Try changing the transformation matrix to one of these and see the different F's you get.

$$\begin{bmatrix} 1 & 0 \\ 0 & -1 \end{bmatrix} \qquad \begin{bmatrix} 0 & -1 \\ -1 & 0 \end{bmatrix} \qquad \begin{bmatrix} -1 & 1 \\ 1 & 1 \end{bmatrix}$$

Rotation matrices

Matrices are used a lot in computer graphics to rotate objects. In 2D, the matrix for rotating an object is

$$R = \begin{bmatrix} \cos\theta & -\sin\theta \\ \sin\theta & \cos\theta \end{bmatrix}$$

where θ is the angle of rotation in degrees. But Python deals with radians, so we need to use the degree converter function we made in the Trigonometry chapter.

```python
def rotate(degrees):
    '''Returns the matrix for rotating a certain angle in degrees'''
    rads = convToRads(degrees)
    #Define Rotation matrix:
    rotmatrix = [[cos(rads),-sin(rads)],
                 [sin(rads),cos(rads)]]
    return rotmatrix
```

Next we'll set the rotation to 45 degrees and multiply the F-matrix by that rotation matrix to get the points for the rotated F:

```python
#Multiply F-matrix by rotation matrix
```

```
#We'll start off testing a 45 degree rotation
rotmat = multMatrix(fmatrix,rotate(45))
```

Now we'll graph it:

```
setup()
drawf(fmatrix) #draw the original F
pensize(2) #makes transformed F thicker
drawf(rotmat) #draw the rotated F
```

You'll have to import a few tools from the math module, and from the trigonometry module you created!

```
from math import sin,cos
from trigonometry import convToRads
```

Now run the module.

Here's the output:	Here's the rotation if you change "degrees" to 120

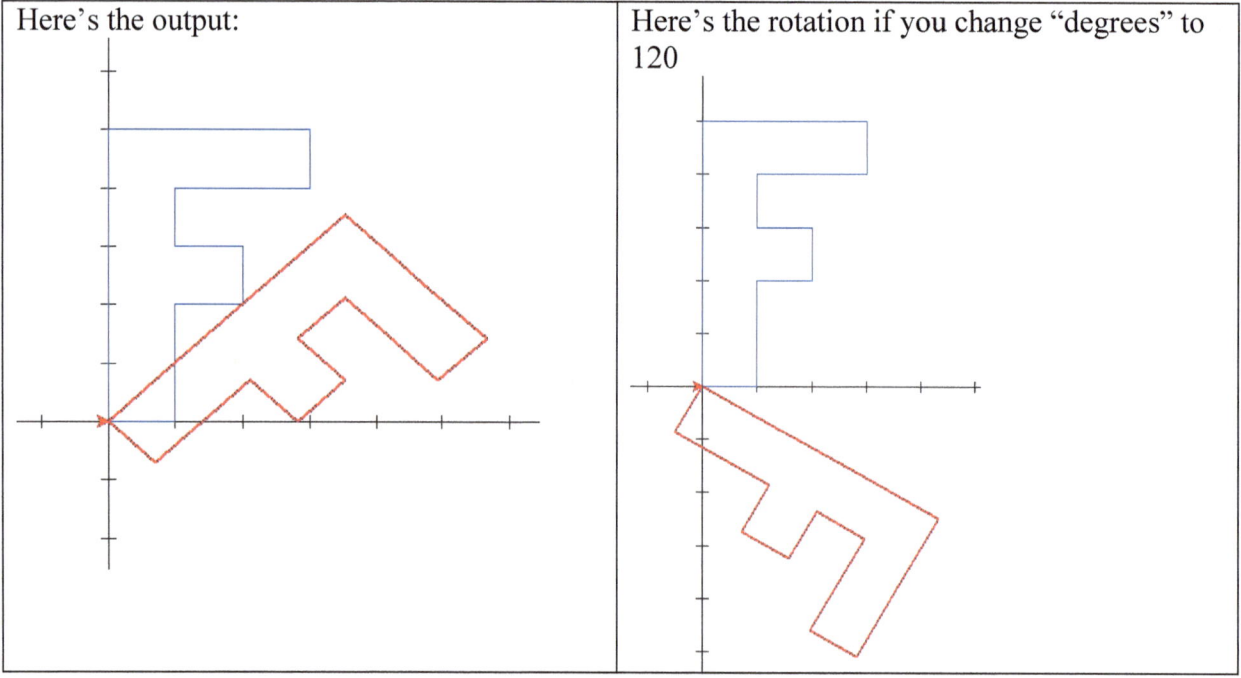

Python has some very powerful tools for working with matrices. **Numpy** is a numerical package used by scientists and others needing lightning fast matrix calculations. The tools we made have been fast enough for our purposes, though.

Matrices Exercises (Solutions on page 142)

Problems 1 and 2 use these matrices:

$$A = \begin{bmatrix} 2 & -3 \\ -1 & 0 \end{bmatrix} \text{ and } B = \begin{bmatrix} 0 & -1 \\ 5 & 7 \end{bmatrix}$$

1. Find AB.
2. Is it the same as the product BA?
3. Find this product:

$$\begin{bmatrix} 5 & 7 \\ -4 & 0 \end{bmatrix} \begin{bmatrix} 9 & -1 \\ 0 & 2 \end{bmatrix}$$

4. $C = \begin{bmatrix} 5 & 7 \\ -4 & 0 \end{bmatrix}$ and $D = \begin{bmatrix} 1 & 0 \\ 0 & 1 \end{bmatrix}$

D is known as the "identity matrix." Multiply A and B to see why.

5. Find out (by multiplying them) why $M = \begin{bmatrix} 2 & 1 \\ 1.5 & 0.5 \end{bmatrix}$ and $N = \begin{bmatrix} -1 & 2 \\ 3 & -4 \end{bmatrix}$ are called "inverses" of each other.

9. Series

Before we could just punch a key on our calculators, it was a real chore to get more and more accurate approximations of non-repeating, non-terminating decimals like pi and e. The most convenient method was infinite series. Just by taking more and more terms of a series like

$$\frac{\pi}{4} = 1 - \frac{1}{3} + \frac{1}{5} - \frac{1}{7} + \ldots$$

you could get as accurate a value of pi as you wanted. At least in theory. The above series, which was known in medieval India centuries before it was "discovered" in Europe, converges to pi so slowly as to be almost useless. Let's take a look using Python:

```python
def piseries(n):
    '''Uses n terms of Gregory's series to approximate pi'''
    running_sum = 0
    for x in range(1,n + 1):
        term = ((-1)**(x - 1))*(1/(2*x - 1)) #calculates one term
        running_sum += term
    print(4*running_sum)
```

After 1000 terms:
```python
>>> piseries(1000)
3.140592653839794
```

Adding up a thousand terms would have taken forever, and we'd only get two correct decimal places? A better series was needed. In the 1700s John Machin came up with this one:

$$\pi/4 = 4 \arctan(1/5) - \arctan(1/239)$$

The arctan series is similar to the one we saw above:

$$arctan(x) = x - \frac{x^3}{3} + \frac{x^5}{5} - \frac{x^7}{7} + \ldots$$

x has to be less than one. That's not a hard Python program to write:

```python
def arctan(x,n):
    '''Returns the inverse tangent of x
    using n terms of series'''
```

```
    running_sum = 0
    for i in range(1,n-1): #starts with 1
        term1 = ((-1)**(i - 1))*(x**(2*i - 1)/(2*i - 1))
        running_sum += term1
    return running_sum
```

Let's test it. According to my calculator, $\tan^{-1}(0.5) = .463647$
```
>>> arctan(.5,20)
0.46364760900064694
```

Now let's calculate pi:

```
def machin(n):
    '''Returns pi using n terms of Machin's series'''
    return 4*(4*arctan(1/5,n) - arctan(1/239,n))
```

```
>>> machin(10)
3.1415926535886025
```

Using only 10 terms, we have around 10 correct digits of pi!

Sums of Infinite Series

Python is useful for finding the sums of infinite series like

$$\tfrac{1}{2} + (\tfrac{1}{2})^2 + (\tfrac{1}{2})^3 + \dots$$

Here's the program:
```
def infSeries(start,ratio,terms):
    '''Returns the sum of a geometric series
    of a given number of terms with given
    starting number and ratio'''
    running_sum = 0
    for i in range(terms):
        running_sum += start*ratio**i
    print(running_sum)
```

To get a sum of the above series after 10 terms, this is what we enter:
```>>> infSeries(.5,.5,10)```
```0.9990234375```
	And after 100 terms:
	```>>> infSeries(.5,.5,100)```
	```1.0```

Now find the sums of these series and see if you can find the pattern:

$$\tfrac{1}{3} + \left(\tfrac{1}{3}\right)^2 + \left(\tfrac{1}{3}\right)^3 + \ldots = ?$$
$$\tfrac{1}{4} + \left(\tfrac{1}{4}\right)^2 + \left(\tfrac{1}{4}\right)^3 + \ldots = ?$$
$$\tfrac{1}{5} + \left(\tfrac{1}{5}\right)^2 + \left(\tfrac{1}{5}\right)^3 + \ldots = ?$$

Iteration

A powerful tool in math is putting an input into a function, getting an output, then putting it back into the function, again and again. I can't believe this method was even considered before computers, but geniuses like Newton and Euler liked it. Now it's not so hard to do using Python.

Program: Square root using Iteration

Newton came up with a way to find accurate approximations of square roots by taking a guess. Call the number you want the square root of x. Take a guess and call that g. Then you average x/g and g and that becomes the new guess. You put the new guess in for g and get yet another guess. And so on. Let's see how that would look in Python:

```python
def newtonRoot():
    '''Calculates Square Root of x by Newton's Method'''
    num = float(input("What number do you need the square root of? "))
    guess = float(input("Enter a first guess. "))
    for i in range(5):
        guess = (num/guess + guess)/2
        print("new guess is", guess)
    print("the final guess is",guess)
    print("that squared is",guess**2)
```

And here's the interactive output. It only takes 5 iterations to get 10 correct decimal places!
```
What number do you need the square root of? 600
Enter a first guess. 20
new guess is 25.0
new guess is 24.5
new guess is 24.4948979592
new guess is 24.4948974278
new guess is 24.4948974278
the final guess is 24.4948974278
that squared is 600.0
```

10. Complex Numbers

Complex numbers are easy to manipulate using lists. You can write 3 + 2i as [3,2] and write functions to add and subtract them.

```
def addComplx(a,b):
    '''Returns the sum of two complex numbers'''
    return [a[0] + b[0],a[1] + b[1]]
```

To add 3 − 4i and 7 + 5i, for example, just enter them as lists:

```
>>> addComplx([3,-4],[7,5])
[10, 1]
```

Multiplying complex numbers is just like FOIL, but since i^2 = -1,

$$(a+bi)*(c+di) = (ac - bd) + (bc+ad)i$$

```
def cMult(a,b):
    '''Returns the product of two complex numbers'''
    return [a[0]*b[0]-a[1]*b[1],a[1]*b[0]+a[0]*b[1]]
```

Run it and multiply two complex numbers:

```
>>> cMult([3,-4],[7,5])
[41, -13]
```

That means the product is 41 − 13i.

Polar Form

Dividing complex numbers is harder. But there is a trick: convert complex numbers of the form a + bi into polar form: **r(cosθ + *i*sinθ)** where r is the magnitude of the vector and theta is the angle of rotation from the x-axis.

Let's graph complex numbers and see what's happening. I've copied and pasted my "mark" and "setup" functions from the Algebra graphers, and now I'll just add:

```
from turtle import *
from algebra import setup
def graph(z):     #The complex number z is entered as a list: []
    speed(0)    #Sets speed to the fastest
```

```
setpos(0,0)
color("black")    #Sets its color
shape("circle")
pd()
st()                    #show turtle
pensize(2)
setpos(z[0],z[1])      #the two elements of list z
clone()              #places a circle on that spot
```

```
>>> z1 = [3,4]
>>> setup()
```

```
>>> graph(z1)
```

The graphs of complex numbers look a lot like regular (x,y) points.

The polar form of this one has an r of $\sqrt{3^2 + 4^2} = 5$ and you find theta by taking the arctangent of 4/3.

```
>>> from math import atan
>>> atan(4/3)
0.9272952180016122
```

That's in radians. If you want a result in degrees, you'll have to convert it yourself. Good thing we created a tool for converting radians to degrees. We'll also import our "root" function for the square root. These are all the imports:

```
from turtle import *
from algebra import setup
from math import atan, sin, cos, sqrt
from trigonometry import convToDegs, convToRads

def rTheta(z):
    '''Converts rectangular form to polar'''
    r = sqrt(z[0]**2 + z[1]**2) #find the hypotenuse
    theta = convToDegs(atan(z[1]/z[0]))
    return [r,theta]
```

Now it'll return the polar form:

```
>>> rTheta([3,4])
[5.0, 53.13010235415598]
```

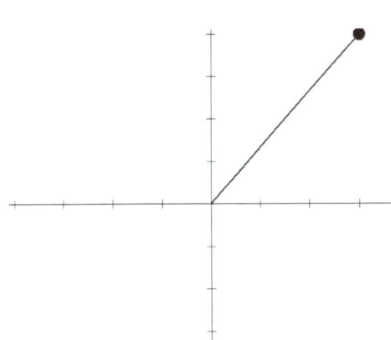

So a + bi can also be written
$r(\cos\theta + i\sin\theta)$, or to save space, "$r(\text{cis}\theta)$."
3 + 4i can be written 5cis(53.1°). That means the length of
3 + 4i is 5 and the angle it makes with the x-axis is around 53
degrees.

This tool works perfectly except for when the "real" part is 0,
like 2i, 3i and so on. We'll just add a conditional for that one
possibility:

```python
def rTheta(z):
    '''Converts rectangular form to polar'''
    if z[0] == 0:
        r = z[1]
        theta = 90.0
    else:
        r = sqrt(z[0]**2 + z[1]**2)
        theta = convToDegs(atan(z[1]/z[0]))
    return [r,theta]
```

```
>>> rTheta([0,1])
[1, 90.0]
```

We did all this to get complex numbers in polar form so we can easily multiply and divide
complex numbers. Multiplying and dividing follow the same pattern:

$$z_1 z_2 = r_1 r_2 (cis(\theta_1 + \theta_2)) \text{ and } \frac{z_1}{z_2} = \frac{r_1}{r_2}(cis(\theta_1 - \theta_2))$$

Now we can make a tool for dividing complex numbers:

```python
def cDivide(z1,z2):
    '''Returns z1 / z2 (rectangular)'''
    new_z1 = rTheta(z1) #Convert to polar
    new_z2 = rTheta(z2)
    new_r = new_z1[0]/new_z2[0] #Divide r's
    new_theta = new_z1[1] - new_z2[1] #subtract thetas
    polar_form = [new_r,new_theta]
    return polar_form
```

This is a problem from a Precalculus textbook:

Convert to polar form and then divide $\frac{1+i}{1-i}$. Express your answer in polar form.

```
>>> cDivide([1,1],[1,-1])
 [1.0, 90.0]
```

The answer is $1\ cis(90°)$.

DeMoivre's Theorem

There's a powerful tool called DeMoivre's theorem, which will make it really easy to take a complex number to an exponent, either as a higher power or a root:

$$z^n = r^n(cis(n\theta))$$

To use this on a complex number in rectangular form, we can use our "r_theta" converter above, but then we'll need to convert it back to rectangular. Here's the code:

```python
def convertRect(z):
    '''Converts polar form to rectangular'''
    r = z[0]
    theta = z[1]
    a = r*cos(convToRads(theta))
    b = r*sin(convToRads(theta))
    return [a,b]
```

Finally, here's our program to take a complex number to an exponent:

```python
def compExpo(z,exponent):
    '''Returns (a + bi) to an exponent'''
    new_z = rtheta(z) #First convert to polar
    new_r = new_z[0]**exponent #Use DeMoivre's Theorem
    new_theta = new_z[1]*exponent
    return convertRect([new_r,new_theta])
```

Here's a surprising question: if **i** is the square root of -1, then what is the square root of **i**?

It's not so hard with DeMoivre's theorem. **i** can be expressed as 0 + 1i or [0,1]. The square root means the exponent is ½.

```
>>> compExpo([0,1],0.5)
[0.7071067811865476, 0.7071067811865475]
```

So the square root of **i** is 0.707 + 0.707*i*.

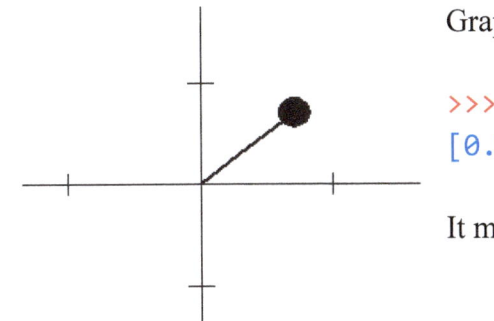

Graph it. What's the polar form?

```
>>> rTheta([.707,.707])
[0.9998489885977782, 45.0]
```

It makes sense that its length is 1 and it's a 45 degree angle.

In this chapter you created a lot of tools for multiplying, dividing and converting complex numbers:

addComplx

cMult

rTheta

cDivide

convertRect

compExpo

Complex Numbers Exercises
(Solutions on page 143)

1. Express 5 – 12i in polar form.

2. Express 8 + 15i in polar form.

3. Multiply (1 – i)(3 – 3i) and express your answer in polar form.

4. Find the cube root of 2 + 11i. (Be sure to use a decimal point.) Rafael Bombelli calculated this in the 1500s long before Python was created!

5. Find the square root of 1 + i and express it in polar form.

6. Find $(1 + i)^{10}$ and express it in rectangular and polar form.

Graphing Complex Multiplication

Let's graph what happens when we multiply complex numbers:

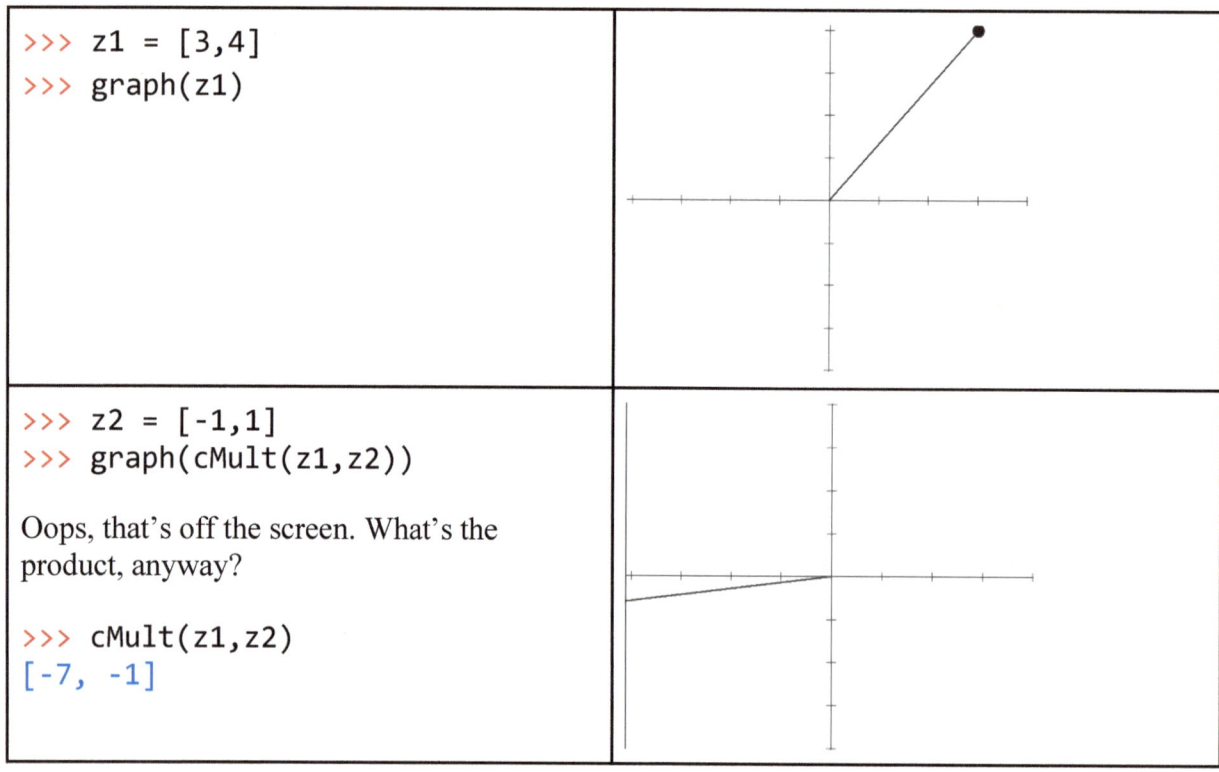

```
>>> z1 = [3,4]
>>> graph(z1)
```

```
>>> z2 = [-1,1]
>>> graph(cMult(z1,z2))
```

Oops, that's off the screen. What's the product, anyway?

```
>>> cMult(z1,z2)
[-7, -1]
```

When you multiply these complex numbers, they get big very quickly.

By making a loop, we can continuously square a complex number and see how it changes position:

```
def graph(z):     #The complex number z is entered as a list: []
    speed(0)     #Sets speed to the fastest
    color("black")  #Sets its color
    shape("circle")
    st()                    #show turtle
    pensize(2)
    setpos(z[0],z[1])     #the two elements of list z
    pd()
    clone()

setup()
c = [0.5,0.75] #initial value of c: 0.5 + 0.75i
for i in range(10):   #Do this 10 times:
    graph(c)          #graph c
    c = cMult(c,c)    #then square c
```

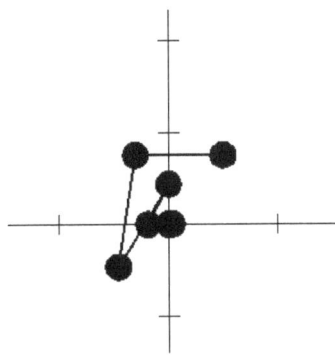

Here's the output:

The first value is on the top right and the subsequent values are closer and closer to the origin. So when iterated, c "converges."

When we put in a different value for c, like [1,1], it diverges:

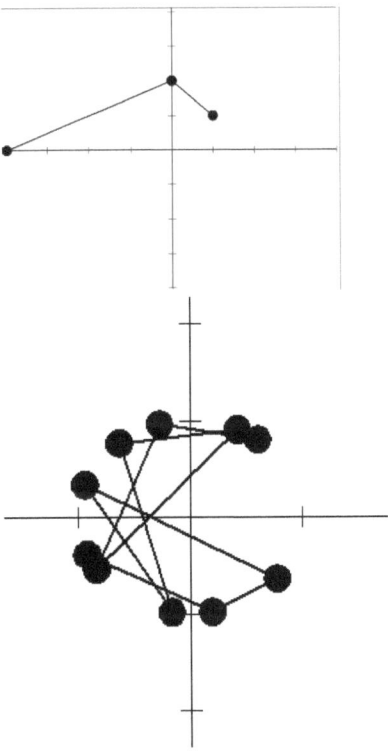

Here the initial value was 0.6 + 0.8i. After 10 iterations, it hasn't diverged.

Mandelbrot Set

It would be great if we could just use a loop and try every point on the grid and put the ones that don't diverge into a set.

In the 1970s a mathematician named Benoit Mandelbrot was using a computer to do just that, and the set of complex numbers that converge is called the Mandelbrot Set. You probably recognize its shape:

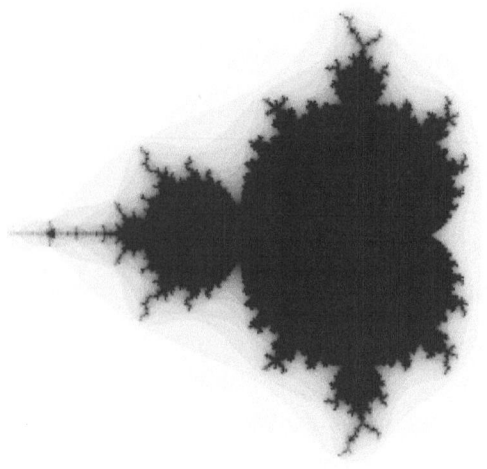

Most math textbooks contain pictures of the Mandelbrot set, but none of them teach you exactly how to create one. That's going to change here.

We're really just doing what we did above: squaring complex numbers and separating the ones that converge from the ones that diverge.

In the first step, all the points that are within 2 units of the origin form a circle, of course. But square them once and some points diverge, leaving an oval.

Square again and more points diverge, leaving a pear shape:

Keep up this process and the points that haven't diverged gets smaller and smaller until it starts to take on the familiar shape.

Starting to look familiar? Another job will be coloring the set so you can see each iteration. Tkinter can easily do color names like "red" and "blue" but if you just want to use numbers for colors (RGB values) you have to convert them to strings. But it'll be worth it!

First we'll import the Tkinter graphics package that comes pre-loaded in Python.
```
from tkinter import *
```

We'll also need the square root function to calculate distance:
```
from math import sqrt
```

This sets up the window, or "canvas:"
```
tk = Tk()
canvas = Canvas(tk,width = 800, height = 800)
canvas.pack()
```

Next we'll define the arithmetic for squaring a complex number z = a + bi

```
def squareZ(z):
    return [z[0]**2-z[1]**2,2*z[0]*z[1]]
```

Here we define the iteration

$$z_{n+1} = z_n^2 + c$$

z starts off as 0 + 0i and is squared and added to the complex number c. That sum becomes the new z and the process is repeated over and over.

```
def mandelbrot(c,iterations):
    z = [0,0]
    for i in range(iterations):
        z2 = squareZ(z) #Square z
        z2[0] += c[0] #Add c
        z2[1] += c[1]
        z = z2 #update z to new value
        #if it diverges on iteration number i:
        if sqrt(z2[0]**2 + z2[1]**2) > 2:
            return i
```

```python
    #If the point hasn't diverged:
    if sqrt(z2[0]**2 + z2[1]**2) < 2:
        return 'black' #The Mandelbrot Set will be colored black
    return i

#Setup the screen:
xlow = -2.5 #change to zoom
dx = 0.005

ylow = -2.0
dy = 0.005
```

Here's the loop to check all the points on the graph:

```python
for i in range(800):
    for j in range(800):
        x = xlow + i*dx #Changes the screen format
        y = ylow + j*dy #to the x-y format
        loc = [x,y]    #The complex form of the current point

'''Now we'll run the mandelbrot function, iterating the complex
coordinates of each pixel on the screen. The function returns the
number of iterations it takes for the point to diverge.'''

        man = mandelbrot(loc,20)
        if man == 'black':
            canvas.create_rectangle(i,j,i+1,j+1,
                        fill = 'black',
                        outline='black')
        else:
'''Tkinter has a special rgb form for its colors.
I made all 3 numbers the same so it'll be gray.'''

            tk_rgb = '#'+str(5*man)+str(5*man)+str(5*man)
            canvas.create_rectangle(i,j,i+1,j+1,
                        fill=tk_rgb,
                        outline=tk_rgb)

mainloop()
```

Run this module and it may take a few minutes but you'll see a Mandelbrot Set. Be patient; the program is taking 640,000 points and iterating each one numerous times to see if and when it diverges!

Tkinter has a lengthy list of available colors if you want to just call them by name. Below is a Mandelbrot drawing program where the colors are put manually into a list. The iteration at which a point diverges will determine the index of the list, and therefore the color it gets. You can search for "Tkinter colors" to get a full list of available colors and choose your own. I'm going with a blue motif:

```python
from tkinter import *
from math import sqrt

#start tkinter and create a drawing canvas
tk = Tk()
canvas = Canvas(tk,width = 800, height = 800)
canvas.pack()

#define the colors we'll be using
fill_color = ['SlateBlue1','SlateBlue2','SlateBlue3', 'SlateBlue4',
              'RoyalBlue1', 'RoyalBlue2','RoyalBlue3',
              'RoyalBlue4','DodgerBlue2','DodgerBlue3', 'DodgerBlue4',
              'SteelBlue1', 'SteelBlue2','SteelBlue3',
              'SteelBlue4','DeepSkyBlue2', 'DeepSkyBlue3',
              'DeepSkyBlue4', 'turquoise1', 'turquoise2','turquoise3']

def squareZ(z):
    '''Define the arithmetic for squaring
    a complex number a + bi'''
    return [z[0]**2-z[1]**2,2*z[0]*z[1]]

def mandel(z,iterations):
    '''Define the iteration zn + 1 = z**2 + c'''
    z2 = z # we'll be playing around with z2
    for i in range(iterations): #"do this a bunch of times"
        z2 = squareZ(z2)        #first square z2
        z2[0] += z[0]           #then add the components of c
        z2[1] += z[1]
        if sqrt(z2[0]**2 + z2[1]**2) > 2: #if the numbers are
            return i #diverging give the iteration number
```

```python
    if sqrt(z2[0]**2 + z2[1]**2) < 2: #otherwise draw black
        return 'black'
    return i


xlow = -2.5 #change to zoom
dx = 0.005

ylow = -2.0
dy = 0.005

for i in range(800): #now we'll loop over all the
    for j in range(800): #pixels on the screen
        x = xlow + i*dx #change the x-y coordinates
        y = ylow + j*dy #to canvas coordinates
        loc = [x,y]     #the current location we're looking at
        iteration = mandel(loc,20) #find the step it diverges at
        if iteration =='black':    #draw a black rectangle
            canvas.create_rectangle(i,j,i+1,j+1,
                        fill='black',
                            outline='black')
        else: #otherwise draw a rectangle in its divergence color
            canvas.create_rectangle(i,j,i+1,j+1,
                        fill=fill_color[iteration],
                            outline=fill_color[iteration])

mainloop() #start the program automatically
```

11. Probability and Statistics

There are plenty of formulas in Probability that could use the Python treatment. The first is the factorial:

$$n! = n\,(n-1)(n-2) \cdot \ldots \cdot (3)(2)(1)$$

We've already seen the way to write a function to return the factorial of a number using recursion in the fractal chapter:

```python
def factorial(n):
    '''Returns n!, the factorial of n'''
    if n == 0: return 1
    return n * factorial(n - 1)
```

Permutations and Combinations

The factorial tool is used in the formulas for the number of permutations and combinations of n objects taken r ways. Here's the permutation formula:

$$nPr = \frac{n!}{(n-r)!}$$

```python
def permutations(n,r):
    '''Returns nPr, the number of ways
    r items can be chosen from a total of
    n, and order matters.'''
    return int(factorial(n)/factorial(n-r))
```

Since order doesn't matter, there are less ways to combine n items r ways. The number of permutations gets divided by the factorial of r:

$$nCr = \frac{n!}{(n-r)!\,r!}$$

```python
def combinations(n,r):
    '''Returns nCr, the number of ways
    r items can be chosen from a total of
    n, if order doesn't matter.'''
    return int(factorial(n)/(factorial(n-r)*factorial (r)))
```

Standard Deviation

Here's the complicated formula for the Standard Deviation of a list of numbers:

$$\sigma = \sqrt{\frac{1}{n}\sum_{i=1}^{n}(x_i - \bar{x})^2}$$

In order to calculate the standard deviation of a group of numbers, you have to
1. Calculate the mean
2. Find the difference of each number from the mean
3. Square the difference
4. Add up all the squares
5. Divide by the number of items in the list
6. Take the square root of the result

This is no small task, but if we do it right, we only have to do it once in Python and we'll never have to go through all that again!

First find the average of the list:

```python
def mean(lista):
    '''returns the average of a list of numbers'''
    return sum(lista)/len(lista)
```

Here's the code for the Standard Deviation:

```python
def stdDev(lista):
    '''returns the standard deviation of a list of numbers'''
    mean1 = mean(lista) #find the mean

    #the running sum of the squares of the differences:
    sumDif = 0

    for x in lista: #find the difference of each term and the mean
        sumDif += (x - mean1)**2 #square it and add it to sumDif

    #divide by the length of the list and take the square root
    stddev = sqrt(sumDif/len(lista))

    return stddev
```

Now you can calculate the Standard Deviation of any list just by entering the numbers:
```python
>>> stdDev([28,84,67,71,92,37,45,32,74,96])
23.992498827758645
```

Probability/Statistics Exercises
(Solutions on page 143)

1.   Calculate 10!

2.   How many ways are there to arrange 6 different books on a shelf? (Order matters.)

3.   How many different teams of 4 players could you make from a group of 20 people?

4.   How many different ways could you arrange 30 people 5 at a time? (Order matters.)

5.   Calculate $_{10}P_8$

6.   Calculate $_{10}C_8$

7.   Calculate the Mean and Standard Deviation of this list of test scores:

78, 67, 92, 91, 62, 94, 98, 51, 72, 81, 61, 85, 88, 42, 78, 78, 80, 78, 68, 82

12. Calculus

Computers should be mandatory tools for learning Calculus, which is the science of change.

Derivatives

The derivative is the slope of a function at a point. Finding derivatives algebraically can be difficult, but finding a numerical value is easy. If you remember slope (rise over run), you can understand derivatives:

Program: Finding the Deriviative at a Point

```python
def f(x):
    return x**2

def derivative(a):
    '''Returns the derivative of x at a point'''
    dx = 0.001   #The tiny change in x
    dy = f(a + dx) - f(a) #The corresponding change in y
    return dy/dx
```

The derivative of $y = x^2$ at $x = 3$ is:
```
>>> derivative(3)
6.000999999999479
```

Newton's Method

In Algebra we learned how to solve equations that have rational roots. As soon as Newton invented Calculus he used it to get accurate approximations of roots of polynomial roots by iteration. The method is based on the idea that the derivative is the slope of a line segment, like in the figure on the left.

AC is the tangent line to the curve at A. The derivative is the slope, which is AB/BC. If we wanted to get closer to the root (just to the left of C in the figure), we could use the equation of line AC in point-slope form:

$$y = f'(B) \cdot (x - B) + f(B)$$

We want where y = 0 and x will just be the next x-value in the process.

So the equation becomes

$$0 = f'(x_n) \cdot (x_{n+1} - x_n) + f(x_n)$$

Rearranging:

$$x_{n+1} = x_n - \frac{f(x_n)}{f'(x_n)}$$

We'll plug in a value for x, do the number crunching above, and put the new value back in for x and so on.

Program: Finding a root using Newton's Method

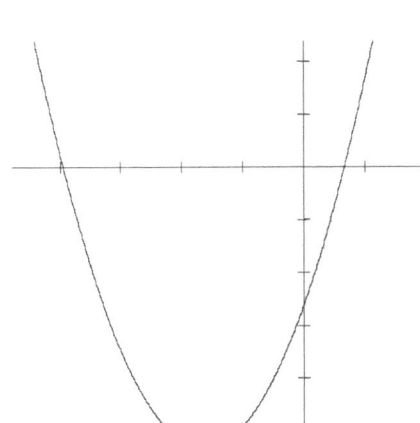

This polynomial is $y = x^2 + 3.3x - 2.6$. Graphing us gives us a rough guide:

It looks like there are two roots: one is around -4 and the other is between 0 and 1. We'll use those approximations as guesses.

```python
def f(x):
    return x**2 + 3.3*x - 2.6
```

```python
def newton(guess):
    for i in range(20):
        new_guess = guess - f(guess)/derivative(guess)
        guess = new_guess
    print(new_guess)
```

This will give us a root of the polynomial:
```
>>> newton(-4)
-3.95705439901
```

That's one. Let's guess x = 1 to find the other:
```
>>> newton(1)
0.657054399012
```

Those are the exact values of the points we saw on the graph. Often we have a higher degree equation like

$$1.6x^5 + 3.9x^4 - 1.9x^3 - 5x^2 + .4x + 0.5 = 0$$

Graphing it, we can see it has 5 roots.

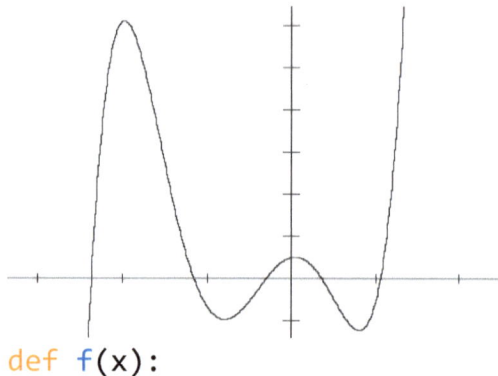

We might not want to have to type in all our guesses, so we'll change our program a little to plug in values:

```python
def f(x):
    return 1.6*x**5 + 3.9*x**4 -1.9*x**3 - 5.0*x**2+.4*x+0.5
```

```python
def newtPoly():
    '''Returns the roots of f(x) by plugging in
    values between -10 and 10'''
    x_plug = -10
    while x_plug <=10:
        x = x_plug #now we'll iterate with x
        for i in range(50):
            new_x = x - f(x)/derivative(x)
            x = new_x #plug the new x back in
        print(new_x)    #print result
        x_plug += 0.5 #plug in next higher x
```

Now the output will contain a lot of repetition (which I've edited out), but we get all five roots:
```
>>> newtPoly()
-2.37451641532
-1.1645498987
-0.303549305456
0.354287480882
1.0508281386
```

Integrals

Integral Calculus is concerned with finding the area under functions. The method usually used is to draw rectangles under the curve, like this:

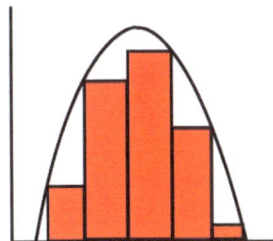 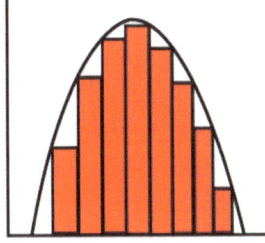 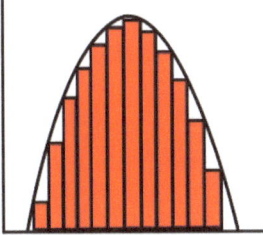

The more rectangles you use, the move accurate the approximation. This is an ideal job for a computer!

Numerical Integration

Program: calculating the numerical Integral under a curve

```python
def f(x):
    return -0.2*x**5 + 1.4*x**4+x**3-5*x**2-1.5*x + 3

def nint(f,startingx, endingx, number_of_rectangles):
    '''returns the area under a function'''
    sum_of_areas = 0 #running sum of the areas
    #width of every rectangle:
    width = (endingx - startingx) / number_of_rectangles
    for i in range(number_of_rectangles): #repeat for every rectangle
        height = f(startingx + i*width) #calculate each height
        area_rect = width * height    #area is width times height
        sum_of_areas += area_rect   #add area to running sum of areas
    return sum_of_areas
```

So to find the area under the curve between x = -1 and 1 using 20 rectangles:
```python
>>> nint(f,-1,1,20)
3.2893239999999997
```

The more rectangles you use, the more exact the approximation:

```python
>>> nint(f,-1,1,200)
3.2335933323999986
```

The Trapezoidal Method

A more accurate way to approximate the integral is to use trapezoids instead of rectangles:

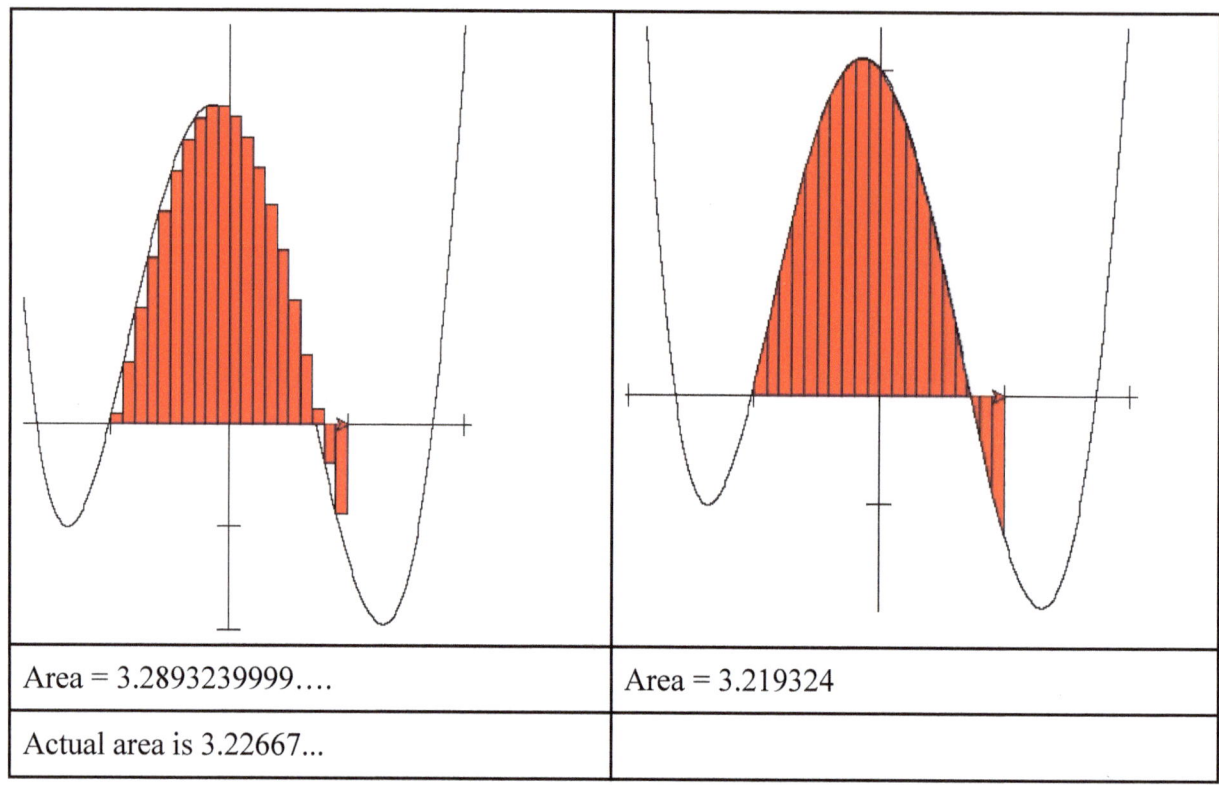

Area = 3.2893239999….	Area = 3.219324
Actual area is 3.22667...	

Program: Finding Area Using the Trapezoidal Method:

```python
def trapezoid(startingx, endingx,numberofTrapezoids):
    '''Returns the area under a function
    using Trapezoidal Method'''
    width = (float(endingx) - float(startingx))/ numberofTrapezoids
    area = 0
    for i in range(numberofTrapezoids):
    #backslash simply continues the code on the next line:
        area1 = 0.5*width*(f(startingx + i*width)+\
                        f((startingx + i*width)+width))
        area += area1
    print(area)
```

This function will get us a more accurate approximation of the area:

```
>>> trapezoid(-1,1,20)
3.2193239999999994
```

Let's graph the Trapezoidal Method. This function draws one trapezoid:

```python
def trap(f,startingx,width):
    pu()              #pen up
    speed(0)             #fastest speed
    setpos(startingx,0) #go to the starting x-coordinate
    setheading(90)       #face straight up
    color('black', 'red')
    pd()                 #put your pen down
    begin_fill()         #start filling in the trapezoid
    height = f(xcor()) #height of the trapezoid
    fd(height)           #go to the top of the trapezoid
    setpos(xcor()+width,f(xcor()+width)) # down the "slant"
    sety(0)              #straight down to the x-axis
    setheading(0)        #face right
    end_fill()           #stop filling the trapezoid
```

We can change our trapezoid function to draw the region and print the area:

```python
from turtle import *
from algebra import setup, graph

def trapezoid2(f,startingx, endingx,numberofTrapezoids):
    '''Calculates area under function f between
    startingx and endingx using trapezoids and graphs it'''
    speed(0)
    setup() #draw the grid
    graph(f) #graph the function
    pu() #pen up. Now calculate the width:
    width = (float(endingx) - float(startingx))/ numberofTrapezoids
    setpos(startingx,0) #go to the starting x-value
    pd()   #get ready to draw!
    area = 0 #running sum of areas
    for i in range(numberofTrapezoids): #repeat for every trapezoid
        trap(f,xcor(),width) #draw a trapezoid. Calculate the area:
        area1 = 0.5*width*(f(startingx + i*width)+f((startingx + \
                            i*width)+width))
        area += area1 #update the running sum of the area
    print(area)
```

Now you can calculate the area using any number of trapezoids you want:

| `>>> trapezoid2(f,-1,1,10)`
`3.197184` | `>>> trapezoid2(f,-1,1,20)`
`3.2193239999999994` |

And the turtles will draw trapezoids in the graph:

10 Trapezoids	20 Trapezoids

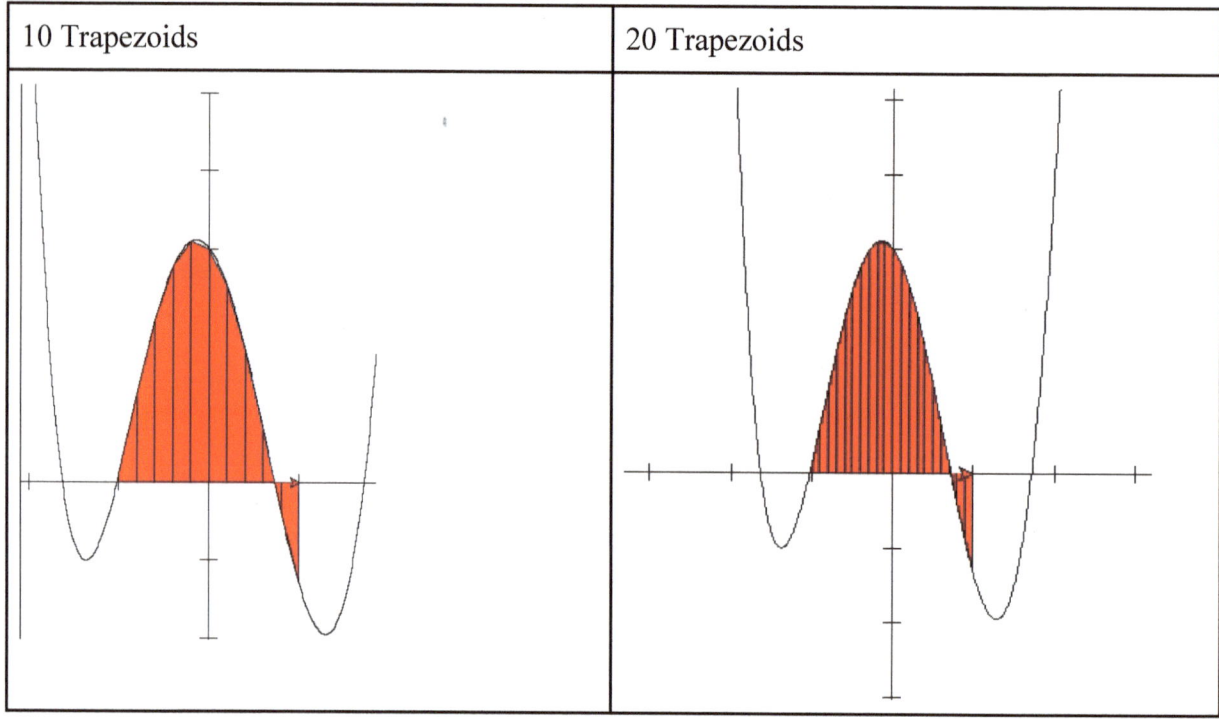

And the area of the 50 trapezoids will be even closer to the actual value of 3.22667:

```
>>> trapezoid2(f,-1,1,50)
3.225493094400001
```

Differential Equations

Differential Equations ("DEs") are a huge topic in higher math, because they come up so often in science, technology and engineering. Simply speaking, they are equations with derivatives in them. For example:

$y' = e^{-y} - y + 1$ means "there is a function whose **derivative** is $e^{-y} - y + 1$."

$y' = x^2 + y^2$ means "there's a function whose **derivative** is $x^2 + y^2$."

A student's task in Differential Equations is often to manipulate these equations algebraically to come up with the mystery function. But most DEs you come across in the real world have no algebraic solution. We have to use Numerical Methods to approximate the solution at whatever value we're interested in. Computers are crucial to this task.

Centuries ago Leonard Euler came up with a method of approximating solutions to DEs by working backwards. For example, if we know one value of a function and we know its derivative, we can start at that value and point in the direction indicated by the derivative. This seems like a good job for our turtles.

From a DE textbook:

Use the Euler method to solve numerically the IVP ("Initial Value Problem")

$$y' = x^2 + y^2$$
$$y(0) = 0$$

on the interval $0 \leq x \leq 1$.

That means we know when x = 0, y = 0 We want to know where it'll be when x = 1. So we'll put a turtle on the point (0, 0).

```python
from turtle import *

def diffeqs():
    '''using turtles to solve differential equations!'''
    setup()
    setpos(0,0) #initial value y(0) = 0
```

Now we know the turtle's direction (or "slope" or "derivative") is the sum of the squares of its x- and y-coordinates, so we'll code that in:

```python
def deriv1():
    return xcor()**2 + ycor()**2
```

Now it takes a little coding to make the turtle face that direction. The turtle understands heading, not slope. Good thing we know that the slope is the tangent of the heading.

```python
def turtleHeading():
    '''converts derivative to heading.'''
    return atan(deriv1())
```

The turtle is starting off at (0, 0), meaning its derivative is

```
>>> deriv1()
0
```

and its heading is

```
>>> turtleHeading()
0.0
```

What if (just to test this) we moved the turtle to (2,3)?

```
>>> setpos(2,3)
>>> deriv1()
13
>>> turtleHeading()
1.4940244355251187
```

Good. The derivative at (2,3) would be $2^2 + 3^2 = 13$, and that corresponds to a heading of 1.494 radians. Heading in turtle geometry should be in degrees, so we'll convert to degrees by importing our convToDegs function from our trigonometry.py file:

```
from turtle import *
from math import pi, atan
from trigonometry import convToDegs

setpos(0,0)

def deriv1():
    return xcor()**2 + ycor()**2

def turtleHeading():
    '''converts derivative to heading.'''
    return convToDegs(atan(deriv1()))
```

Now it will return the heading in degrees:

```
>>> setpos(2,3)
>>> turtle_heading()
85.60129464500447
```

Now we want to have our turtle face the direction indicated by the DE, then go forward a little bit, recalculate its direction, move a little, and so on until its x-coordinate is 1. I imported the

```
```

setup function from the grapher I used in the Algebra chapter.

```python
from algebra import setup

def diffeqs():
    '''using turtles to solve differential equations!'''
    setup()
    setpos(0,0) #initial value y(0) = 0
    pd()
    while xcor() <= 1:
        setheading(turtleHeading())
        fd(0.1)
```

The output looks like this:

So the turtle walked around and stopped when its x-coordinate reached 1. If we want the numbers at every step, we just have to add in a print statement into the while loop:

```python
print(xcor(),ycor())
```

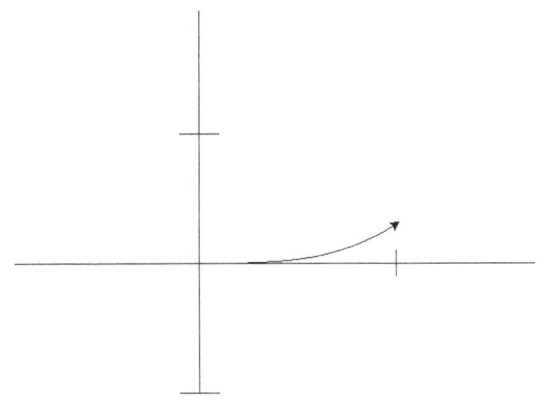

Here's what our printed output looks like:
```
0 0
0.1 5.08560541184e-14
0.199995000375 0.000999950003851
0.299915100237 0.00499665408576
0.399512774973 0.013957858649
0.49825983609 0.0297381603194
0.595293524998 0.0539138343937
0.689463592517 0.0875590180509
0.779509305689 0.131053495164
0.864316385777 0.184041787899
0.943131661513 0.245589755458
1.015638423545 0.314457523387
```

That means when x = 1, the y-value is around 0.3. That's what we wanted to know!

The Runge Kutta Method

The Euler Method was eventually improved on, and a widely used method is the Runge-Kutta Method, defined by these complicated-looking formulas:

$$l_1 = hf(x_i, y_i)$$
$$l_2 = hf(x_i + \frac{1}{2}h, y_i + \frac{1}{2}l_1)$$
$$l_3 = hf(x_i + \frac{1}{2}h, y_i + \frac{1}{2}l_2)$$
$$l_4 = hf(x_i + h, y_i + l_3)$$
$$y_{i+1} = y_i + \frac{1}{6}(l_1 + 2l_2 + 2l_3 + l_4)$$

But you only have to code it in Python once. I won't use lower-case l's because they look too much like ones:

```python
#Runge-Kutta Method for solving DEs

def deriv(x,y):
    return x**2 + y**2

def rk4(x0,y0,h): #order 4
    while x0 <= 1.0:
        print(x0,y0)
        # I changed the l's to m's
        m1 = h*deriv(x0, y0)
        m2 = h*deriv(x0 + h/2, y0 + m1/2)
        m3 = h*deriv(x0 + h/2, y0 + m2/2)
        m4 = h*deriv(x0 + h, y0 + m3)
        #These are the values that are fed back into the function:
        y0 = y0 + (1/6)*(m1 + 2*m2 + 2*m3 + m4)
        x0 = x0 + h
```

The ouput for h = 0.2 is

```python
>>> rk4(0,0,0.2)
0 0
0.2 0.00266686669333
```

```
0.4 0.0213600903815
0.6 0.0724512003541
0.8 0.174090180973
1.0 0.350257549145
```

Seems the value at x = 1 may be closer to 0.35. That may have seemed like a lot of work, but now that we have our Runge-Kutta formulas in Python code, we can solve new questions almost instantly:

Use the Runge-Kutta approximation of order 4 with h = 0.2 to solve the IVP

$$y' = y^2 - 2x$$
$$y(0) = 0$$

on the interval $0 \leq x \leq 1$.

Solution: All we have to do is change our "deriv" formula to match our DE:

```python
def deriv(x,y):
    return y**2 - 2*x
```

and run it, and here are our results:

```
>>> rk4(0,0,0.2)
0 0
0.2 -0.03992021312
0.4 -0.157954224581
0.6 -0.345206732286
0.8 -0.581850685324
1.0 -0.839828069894
```

Those values are the approximations of the points in the function over the interval from 0 to 1.

Calculus Exercises:
(Solutions on pages 143 and 144)

For problems 1 - 3, $f(x) = x^2 - 13x + 40$

1. Find the derivative of f(x) at x = 3.5

2. What is the derivative of f(x) at x = 10.25?

3. Find the area under f(x) from x = 3 to x = 5 using 10 rectangles. Compare the area found by using 10 trapezoids. Which is closer to the actual area of 8.667?

4. Find the area under the curve $y = -0.25x^3 + 3x^2 - x - 10$ from x = 3 to x = 4

5. Here is a graph of the function $g(x) = -0.5(x - 2)^2 + 4$. Find the area below the curve and above the x-axis. You'll need to find the intersection points (I didn't include the ticks!) and use integration.

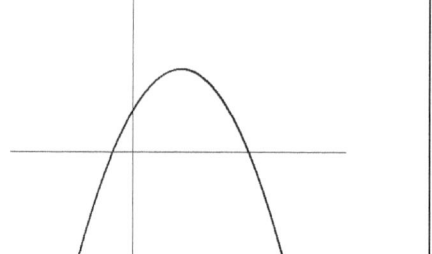

6. Use Newton's Method to find all the solutions to the equation

$$0.1x^5 + 0.6x^4 + 0.6x^3 - 0.7x^2 - 0.6x + 0.3 = 0$$

7. The rate at which a certain object cools is found to be

$$y' = -0.063(y - 20)$$

in degrees per hour. The object starts at a temperature of 37 degrees (meaning the Initial Value is y(0) = 37). Use the Euler or Runge-Kutta method to estimate when the object will be 30 degrees.

Conclusion

We've learned the basics of Python and how to use a handful of Python's ready-made tools like loops, variables, functions, conditionals and lists. With those tools, we made a ton of tools to use for exploring math topics, solving equations, and graphing functions. We even built on those tools, and we'll always have them in our toolbox.

Hopefully the ideas in this book have given you an idea of the possibilities that exist for using computer programming for exploring and visualizing ideas in math from arithmetic to calculus and beyond. It's my hope that you're inspired to go further. I know I will!

Answers to Exercises

Answers to Arithmetic Exercises

1.  ```
    >>> average(225,723)
    474.0
    ```
2.  ```
    >>> average(1412,36877)
    19144.5
    ```
3.  ```
    >>> ctof(25)
    77.0 F
    ```
4.  ```
    >>> ftoc(212)
    100.0 C
    ```
5.  ```
    >>> median([18,12,11.5,14,9,21,8,15,3,25,10,18,6])
    12
    ```
6.  ```
    >>> median([76,59,64,23,11,98,56,77,91,89,48,101,55,37])
    61.5
    ```

7.  ```
    fibo(20)
    [1, 1, 2, 3, 5, 8, 13, 21, 34, 55, 89, 144, 233, 377, 610, 987,
    1597, 2584, 4181, 6765]
    ```
    So 6,765 is the 20th Fibonacci number.

Answers to Algebra Exercises

1. 13
2. -9
3. 8 and 5
4.  ```
    >>> quad(5,-25,-29)
    5.97131099154 , -0.971310991542
    ```
5. x = 3/2, -7, -⅙
6. x = 3/7, -4,-⅕, ⅔
7. x = -1.4, 2/9, .4286, 2, 0.5 +/- 2.5i
8. x = 1.618, -0.618, 0.5, -⅔, -4/9, 8/7, 11
9.  ```
    >>> binary(44)
    101100
    ```
10. ```
    >>> isPrime(1000001)
    1000001 = 101 x 9901.0
    ```
    No, it's not prime.

Answers to Geometry Exercises

1. a. 7.21
   b. y = 1.5x + 3.5
   c. (-1, 2)
   d. y = -⅔ x + 1.33
2. 5.37
3. (0.5, 1.5)
4. a. 21.5
   b. y = 0.625x + 0.5
   c. (0.407, -0.058), radius = 4.718
5. a. AB: y = 0.16x+4.491
     CD: y = -0.013x - 2.187
   b. (-38.437, -1.676)
   c. y = -6.233x + 27.355
6. y = 5x - 33
7. 8.497

Worked Solutions to Geometry Exercises:

```
1.
>>> A = (1,5)
>>> B = (-3,-1)
>>> distance(A,B)
7.211102550927978
>>> line2points(A,B)
(1.5, 3.5)
>>> midpt(A,B)
(-1.0, 2.0)
>>> perpBisect(A,B)
(-0.6666666666666666, 1.3333333333333335)

2.
>>> distPointLine((4,1),2,5)
5.366563145999495

3.
>>> equidPt((-2,6),(5,4),(-4,-1))
(0.5, 1.5)

4.
>>> A = (4,3)
>>> B = (3,-4)
>>> C = (-2,4)
>>> heronPoints(A,B,C)
21.499999999999986
>>> line2points(B,C)
(-1.6, 0.8000000000000007)
>>> perpendLine(-1.6,A)
(0.625, 0.5)
>>> circumcircle(A,B,C)
(0.40697674418604657, -0.058139534883720936)
4.71826594541
```

```
5.
>>> A = (-1.5,4.25)
>>> B = (5.25,5.333)
>>> C = (-4.66, -2.125)
>>> D = (4.75, -2.25)
>>> line2points(A,B)
(0.16044444444444447, 4.490666666666667)
>>> line2points(C,D)
(-0.013283740701381, -2.18690223166843)
>>> perpendLine(.1604,D)
(-6.234413965087282, 27.36346633416459)
>>> intersection(.1604,4.49,-0.013,-2.187)
(-38.50634371395617, -1.6864175317185692)

6.
>>> line(5,(8,7))
(5, -33)

7.
>>> distPointLine((5,-4),2,5)
8.497058314499201
```

Answers to Trigonometry Exercises

```
1.
>>> convToRads(50)
0.8726646259971648

2.
>>> convToDegs(1)
57.29577951308232

3.
>>> lawofCos_c(20,27,34)
15.285268124588297
>>> lawofCos_c(20,17,34)
11.191713481972798
>>> solve_tri_3sides(20,17,11.19)
a =  20
b =  17
c =  11.19
A =  87.8586127644
B =  58.1471662984
C =  33.9942209372
```

```
4.
>>> solve_tri_3sides(6,7,9)
a =  6
b =  7
c =  9
A =  41.752205202
B =  50.9771974348
C =  87.2705973632
```

Answers to Matrices Exercises

```
1.>>> A = [[2,-3],[-1,0]]
>>> B = [[0,-1],[5,7]]
>>> multMatrix(A,B)
[[-15, -23], [0, 1]]

2.>>> multMatrix(B,A)
[[1, 0], [3, -15]]
AB doesn't equal BA

3. >>> C = [[5,7],[-4,0]]
>>> D = [[9,-1],[0,2]]
>>> multMatrix(C,D)
[[45, 9], [-36, 4]]

4. >>> multMatrix(C,[[1,0],[0,1]])
[[5, 7], [-4, 0]]
```

When you multiply C by the identity matrix you get C. It's like multiplying a number by 1.

```
5. >>> M = [[2,1],[1.5,0.5]]
>>> N = [[-1,2],[3,-4]]
>>> multMatrix(M,N)
[[1, 0], [0.0, 1.0]]
>>> multMatrix(N,M)
[[1.0, 0.0], [0.0, 1.0]]
```

When you multiply MxN or NxM you get the identity matrix.

When you multiply a number by its inverse, like 2 times ½, you get 1.

Answers to Complex Numbers Exercises

1.
```
>>> rTheta([5,-12])
[13.0, -67.38013505195958]
```

2.
```
>>> rTheta([8,15])
[17.0, 61.92751306414704]
```

3.
```
>>> rTheta(cMult([1,-1],[3,-3]))
[-6, 90.0]
```

4.
```
>>> compExpo([2,11],1./3.)
[2.0, 1.0000000000000002]
```
The answer is 2 + i

5.
```
>>> rTheta(compExpo([1,1],.5))
[1.189207115002721, 22.5000000000]
```
6.
```
>>> compExpo([1,1],10.0)
[9.797174393178832e-15,
32.00000000000002]
```
"e-15" is scientific notation. It means the decimal point is moved 15 places to the left. So it's 0 + 32i or just 32i.
```
>>> rTheta(compExpo([1,1],10.0))
[32.00000000000002,
89.99999999999999]
```

Answers to Probability/Statistics Exercises

1.
```
>>> factorial(10)
3628800
```

2.
```
>>> factorial(6)
720
```

3.
```
>>> combinations(10,4)
210
```

4.
```
>>> permutations(30,5)
17100720
```

5.
```
>>> permutations(10,8)
1814400
```

6.
```
>>> combinations(10,8)
45
```

7.
```
>>> a = [78,67,92,91,62,94,
98,51,72,81,61,85,88,42,78,78,
80,78,68,82]
```
```
>>> mean(a)
76.3
```
```
>>> stdDev(a)
14.085808461000738
```

Answers to Calculus Exercises

1.
```
>>> def f(x): return x**2-13*x+40
>>> deriv(3.5)
  -5.999000000002752
```

2.
```
>>> deriv(10.25)
7.500999999976443
```

3.
```
>>> trapezoid(3,5,10)
8.68
```

4.
```
>>> trapezoid(3,4,20)
12.56265625
```

5.
```
>>> newtPoly()
-0.828427124746
4.82842712475
```

Those are the intersections. Now find the area under the curve between those values.
```
>>> trapezoid(-.8284,4.8284,20)
15.0472333865
```

6. x = -4.32, -1.78, -1.15, 0.4, 0.84

7.
```
def deriv(x,y):
    return -.063*(y-20)

def rk4(x0,y0,h):          #order 4
    while y0 > 30.0:       #stop when the y-value gets to 30
        print(x0,y0)
        m1 = h* deriv(x0,y0)
        m2 = h* deriv(x0 + h/2., y0 + m1/2.)
        m3 = h* deriv(x0 + h/2., y0 + m2/2.)
        m4 = h* deriv(x0 + h, y0 + m3)
         #These are the values that are fed back into the function:
        y0 = y0 + (1./6)*(m1 + 2*m2 + 2*m3 + m4)
        x0 = x0 + h

>>> rk4(0,37,0.1)
0 37
0.1   36.8932366576
0.2   36.7871438101
...   ...
8.4 30.0142927151
```
So it'll take 8.4 hours to get to 30 degrees.